PAGANS PROGRESS

A GE-OGRAPHY
PRIMER

MICHAEL DAMES

ILLUSTRATIONS BY
NATALIE KAY-THATCHER

Pagans Progress: A Ge-Ography Primer by Michael Dames
First Published by Strange Attractor Press 2017
ISBN: 978-1907222498

Text Copyright © 2017 Michael Dames
Illustrations copyright © 2017 Natalie Kay-Thatcher
Cover design by Tihana Šare

Strange Attractor Press
BM SAP, London, WC1N 3XX, UK
www.strangeattractor.co.uk

Distributed by The MIT Press, Cambridge, Massachusetts.
And London, England.

Printed and bound in the UK by TJ International, Padstow.

Contents

Michael Dames, acting The Fool at Avebury.

INTRODUCTION: A DEATHBED CONFESSION

On completion of a three-year long degree course in geography at Birmingham University, from 1955–1958, I literally did not know the first thing about that subject: namely that it was titled after the ancient Greek goddess Ge, alias Gaia or Mother Earth. But she was not mentioned, either by my lecturers, or by any fellow student.

Our subject was not regarded as a pre-Christian adoration of planet Earth – a divine gift from the goddess, named in Hesiod's *Theogony*, written in the 7th Century B.C. Hesiod described Ge as the first of all the deities. So defined, and underpinned by the same deity's rocky underworld of Ge-ology, Geography was properly an exercise in spiritual appreciation, involving a desire to preserve her gift of the entire world as a sacred, living entity, from the harm that now immanently threatens her well-being.

Yet today Ge's 'past' is still far from over. Following the cyclical, sun-based pattern of the seasons, our farmers, in essence, continue to reiterate the agrarian year, begun more than 6,000 years ago in Ge's Neolithic era. Meanwhile, despite much Christian discouragement, the blurred remains of seasonal folk festivals may still combine with our topography (Ge's Ge-omorphology), to evoke the pregnant deity, recognised in the forms of our hills and river sources and valleys, that continue to inspire artists and poets, who irrepressibly give fresh recognition to her overlooked truths.

This text is a belated attempt to express gratitude for the Ge that I previously failed to recognise, and to offer an alternative to the cult of detached estrangement and disenchantment, upon which I was reared. By doing so, I hope to enhance the sense of connection that other people might enjoy with these islands.

The Ge-Union

The Goddess Ge, named after the Greek word for earth, was recognised by a variety of names, right across the Eurasian continents during the long Neolithic era. Her fundamental importance continued to be acknowledged throughout the Bronze Age, and on into Roman times, when Statius describes how 'altars were wreathed with living, well-grown trees in her honour, as countless flowers, heaps of fruit and new produce was poured untouched upon her altars'. The Goddess is then invoked, 'O eternal creatress of gods and men, who brings into being the seeds of life throughout the world including its seas and airs. You are the firm, steadfast, strength of the entire unfailing universe.'[1] Accompanied by devotion of this range and intensity, the Romans had sufficient encouragement to gather around Ge's manifestation at Silbury, which they clearly did.

In Hesiod's *Theogony*, written in Greece, c. 700 B.C., he also nominates Ge, alias Gaia, as the Universe maker. After the nothingness of the yawning gap (termed Chaos) it was wide-bosomed Earth, who, without a mate, gave birth to starry Heaven (Uranus), to cover her on every side. Then she gave birth to the mountains (Ourea) and to Pontus of the raging oceans. All humanity, along with every other life form, was generated by her. Ge was acknowledged as *the* primordial element, mother of the entire cosmos, and of all the deities imagined within it.[2]

Ge was also the source of the vapours of divine inspiration, and was recognised as a powerful prophetess, and giver of dreams. As the all-producing, nourishing mother, she presided over marriage and birth ceremonies. Her worship was universal among the Greeks, and she had temples in Athens, Sparta, Delphi, Olympia, Bura, Tegea, and Phylus.[3]

In Rome, each newborn baby was normally placed on the ground, in recognition of Earth's gift. In that city Ge became Tellus, the Goddess named from Latin *terra*, ('earth'), while her infernal link to Dis and the Manes was emphasised. Yet her First Fruits festival remained strong. Her *Fordicidia* was celebrated on April 15th, (a day after that of Ceres' feast). At the Fordicidia, a pregnant cow was sacrificed in each of Rome's 30 wards, and another on the Capitol, with the calves being burnt as offerings to the country's flocks, herds and fields.[4] This Roman festival stems from the Greek *Thargelia*, held in the month of May, when the *Thargelos*, or 'First Loaf', made from the 'first fruits' of the new harvest, was ceremonially carried home. By contrast, in colder England, its equivalent, the 'Lammas loaf', did not arrive until August 1st.[5]

Ge's Silbury monument, around which the Romans built a substantial settlement, shares in a tradition of worship that pre-dated their Empire by thousands of years, yet which persisted both in Classical culture, and in that of the natives of this island. The two peoples were both gratefully appreciating the one and only Earth that they, and we, can enjoy in common.[6]

Today, Ge's abiding legacy invites us to re-enter her Sacred Geography, the very prospect of which, so long ago discarded, remains tantalisingly visible, and ready to challenge the dreary statistical account of the world's glories, imposed by the 1960s version of geography. Through the Ge persona, in her original nature, and with the help of our archaeological remains, we may recover our sense of a living place underlying current abstractions. From a starting point around Silbury Hill and its moat, (seen as a gigantic image of the world's loving landlady personified,) we may re-enter the island-wide lost landscapes.

Inside these lost landscapes, networks of meaning combined with natural features and sympathetic architecture can help us towards a new language of inhabitable symbols, perhaps of inexhaustible value. Ge-ography, as a form of religious art, may help us rediscover a healthy future for our planet.

Though the Silbury Hill builders did not speak a word of Greek, they probably shared the sense of earthly good fortune that was expressed in the anonymous *Hymn to Gaia*, written in Greece between the 7th and 4th centuries B.C.:

> I will sing of well-founded earth, mother of all, eldest of all beings. Through you, O Queen, men are blessed in their children, and blessed in their harvests. Mother Earth, source of all gods and men, all nourisher, all giver, all destroyer... Foundation of the endless cosmos, around you heaven's teeming stars revolve in ever-winding flow most awesome. Many splendored lady, with the pains of childbirth you bring forth all manner of life, eternal, revered, deep-bosomed, blessed one.[7]

Considering the range and scale of her gifts, it was surely reasonable for the people of Neolithic Wiltshire to say 'Thank You' to Mother Earth, by constructing their own personification of that world-wide deity. Being illiterate, they chose to embody her not in writing but in architectural form. Their Hill remains the tallest prehistoric structure in Europe.

Not the least of Ge's gifts were (and are) the relationships that she established with her offspring, Uranus, the Sky, and Pontus, the Oceans. Thus planet Earth was connected to sources of heat, light and moisture, in an elemental, fertile union.

I.
Life World

CERIDWEN'S LEGACY
The sacred unity between river, land and ocean.

With the arrival of Christianity in Britain, circa 1,500 years ago, divinity was officially divorced from natural phenomena. Yet, some rivers, such as the river Dee in Wales, retained and extended their sacred value.

The term Dee could be applied to cover innumerable female deities, but in a seminal essay, Professor William J. Gruffydd established that an earlier form of the river's Welsh name, in use during the Middle Ages, was Dyfrdonwy, 'the waters of the goddess Don'.[1] And as Professor John Rhys has shown, Don was the mother of the entire Welsh pantheon, including Ceridwen, the wizard Gwydion, and Amaethon, the prototype husbandman. It was Don who determined the fate of nations by her every watery twist and turn.[2]

Thus in Welsh folk tradition, an additional layer of sanctity was present in the Dee, and in Llyn Tegid, Wales' largest lake, lying close to the river Dee's source. In and around Llyn Tegid lived Ceridwen, whose name has been interpreted to mean either 'the dearly beloved white one' or 'the hunchbacked witch or hag'. She has been seen as 'a Nature goddess', or as 'an old human magician, skilled in the arts of divination and enchantment'. In truth, she crossed many boundaries.[3]

The early medieval tale *Hanes Taliesin*, 'The Story of Taliesin', tells of Ceridwen's relationship with her small male helper, the demi-god called Gwion Bach, (Little Gwion), alias Taliesin. (In north Britain, Taliesin had been a mortal court poet, but after the Brythonic tribes migrated from central Scotland to Wales, his persona took on a divine role, so illustrating the inherent proximity of poets to the sacred.)[4]

At Llyn Tegid, he encountered Ceridwen, as she boiled her annual brew, made from charm-bearing herbs that she had put in a cauldron over a fire, which had to be stirred continually for a year and a day. Then three drops, distilled from that liquor, would contain the combined virtues of all these herbs. She intended to anoint her ugly son named Afagddu-Morfran with the drops, so to invest in him their potent

Ceridwen with her cauldon at Llyn Tegid.

wisdom, which included the power to transmit extraordinary learning in several arts, along with the spirit of prophecy.[5]

But by mischance, the three drops splashed instead onto Gwion Bach, who, on sucking his finger, gained all that wisdom. Sensing Ceridwen's likely rage at this outcome, Gwion immediately rushed into the Dee valley, with her in hot pursuit. In order to evade her wrath, he adopted a number of different animal disguises, becoming in quick succession a fish, a crow, a roebuck and a squirrel, but with each of these metamorphoses, Ceridwen countered by becoming a devouring enemy to devour each of his new forms. At last, exhausted, Gwion became a grain of wheat, whereupon his chaser turned herself into a hen, and swallowed him whole.

The accidental effect of this escapade was to couple Gwion's recently acquired plant-derived wisdom to a wide range of beings in the animal kingdom, thereby further adding to his powers; a process confirmed by the sacred Dee waters, in which the changes had happened. Then as a grain, he enters the womb of the river Goddess in her Ceridwen manifestation. Wheat, even a single grain, can symbolise our most highly valued domesticated crop, grown in Britain since 4000 B.C. A Goddess-given gift and 'child' of the land deity, wheat was eventually celebrated at and on Silbury Hill, as the most important divine transaction between Nature and Agriculture.

At the Dee estuary, Ceridwen decides to regurgitate Gwion, and in his human form. She then puts him into a bag, and throws him into the sea; in which to submerge him in a new task, whereby he carries a heavy responsibility for the continuation of life, animal and vegetable, wild and cultivated, throughout Wales. So he bobs along that country's north coast, imprisoned and in the dark, both physically and mentally. Through this suffering, he completes his radical education into the unity of sacred river, land and ocean.

On May Eve, the start of the summer half of the Celtic year, he is caught in a salmon fishing trap, near Aberystwyth. Pulled from the sea, he is instantly recognised as the returning sun god, and renamed Taliesin, 'Shining Brow', ready to bring the sun's warmth, to match Ceridwen's cauldron fire. By river, land, sea, sun and air, his elemental incorporation is complete, and he can pass this cohesion in a vivid narrative to the entire community, enabling them to retain the sense of cohesion that they had inherited from Antiquity. The 'on' ending of Gwion's name is regarded by linguists as a sure marker of his supernatural function, even if he had stumbled into it by chance.

Britain contains many and varied examples of voyages into the sacred, and in this instance, an odd pair of male and female participants improbably succeed in re-establishing a fragile sense of sun-blessed overall working unity.

Many thanks to Wales, for preserving this simple narrative, that encapsulates a dynamic of energy, size and scope, to form a bridge between her Palaeolithic (Old Stone Age) persona, and Gwion, representing Neolithic agriculture.

SALMON-WOMAN AND THE CYCLE OF LIFE
Ge is the named mother of her own rocks, as detailed in Ge-ology. She is as present in her underworld as on the surface.

Like Ge, Ceridwen, the powerful Welsh goddess-hag of wisdom, rebirth and transformation, is all pervasive. Ceridwen has the ability to change herself into different creatures. She is at home in the water as she is on land.

With Ceridwen intent on keeping the door wide open to the natural world as the divine entity that she personified, it is to be expected that many of her animal and vegetable manifestations in recent times were met on sacred terms. This example, *A Fisher-Story*, collected by Edwin Sidney Hartland, c. 1900 A.D., from a farmer's son, near Llanelly, 'was spoken with dramatic force'.[1]

'Upon the river Towy (in Carmarthenshire) floated a fisher-lad. He was in the very dew of youth. He sat in a coracle, with his paddle stuck under his left arm pit, with his salmon rod and his knocker to kill the fish, all ready. Suddenly a great salmon leapt to his fly, and then there was a long fight, in which at last he got the better, and the big fish was flapping between his feet, with the hook through its upper jaw, on the left. He took his club and said: "now I will knock thee", when the fish reared itself against his leg, and spoke with a faint human voice, as if it was the voice of a baby and said: "no, do not knock me; be my *cariad*, (lover), and I will be thine". "No", said he, trembling with amazement. "Thou art a devil, and I will knock thee", raising his arm to strike.

'But before the blow could fall, he found himself in the arms of a beautiful girl, but cold and wet, who knelt between his feet, her face against his, and her eyes were asking him; "Be my *cariad*". "No" said he. "I will knock thee". "Then I will drown thee," said she, bending him over with all her strength, so they capsized.

'Then the girl plunged him deep in the river, and brought him up spluttering, for he could not swim. Again he refused. "And so down you go again (*yn ngwaelod yr avon*), to the bottom of the river" said she - so down they went. Up again she brought him panting. "No", he said "by----". The word was drowned in his mouth. So she forced him into the weeds at the bottom, then plucking up the all-but drowned lad, he finally said "Yea".

'At that she was delighted, and wrung him in her arms and swam with him to the shore. And the coracle went down stream, and the rod too, but that was held to her by the hook and line, for the hook was still in her upper lip.

'So when he came to his strength, he had with him a girl without a stitch of clothes on her - oh a beautiful girl, white as a salmon, and he trying to get the hook from her lip, but he could not pull it through. "I must cut thee" he said, and took out his little knife. "Yes" said she. "Cut me", and he cut the hook out carefully, and she did not wince,

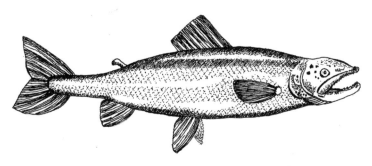

The Salmon, regarded as deliverer of wisdom in Celtic myth.

but kissed him suddenly on the mouth, so that her blood was upon his face.

'"Now thou hast taken of my blood, you will love me for ever" she said, and at that word there came a violent love for her, which never left him during his whole life. He took her home and lived with her a long and lucky life, having many children, who all had a little scar in their upper lips, to the left.' The informant added that if the man had completed his oath, "By God", she would have become *eog*, a salmon again.

On both sides of the Irish Sea, the salmon was for long regarded as a sacred fish, even if that quality does not appear in the Welsh episode described above. Yet if, as is said, folklore represents the debris of paganism, such fragments may retain their vitality and a tenuous connection to their mythic Welsh-Irish origin.

In her salmon form, the goddess Sinnan was the embodiment of the River Shannon, Ireland's longest river. She was required to swim to its submarine spring that lay in the Other World; and there she ate the magic hazel nuts in Connle's submarine well, from which she acquired the bubbles of knowledge and inspiration. Later, when Finn was roasting a salmon, he sucked his blistered thumb, and so received some of the fish's wisdom (reminiscent of Ceridwen's servant, Gwion Bach, accidentally sucking his thumb and gaining shape-shifting powers). As fish and fish-woman, the salmon in Irish accounts was a supernatural being.[2]

The salmon's regenerative role in Irish myth is clearly defined after the mythic Second Battle of Moytirra, during which the human population of Ireland was decimated. According to Gray, the disaster 'illuminated the principals fundamental to the ordering and maintenance of human society, including the contractual and magical aspects of sovereignty.' In this process the salmon played a central role.[3]

Many Irish myths imply that the achievement of good order involves the sacrifice of humanity's separateness from the rest of the living world. Only by empathy with – and in a sense *becoming* – the lower species, can spiritual and physical doom be avoided.

So in this case, the only human survivor, Tuan (whose name means 'wholeness'), spent 300 years as an ox, 200 years as a wild stallion, 300 as a solitary bird, and 100 years as a salmon. In his salmon shape, a fisherman took him to the queen of Lord Redneck, the King of Ulster. The queen wanted the salmon to mate with her. And 'so he was urged of her', and of her, as a new man 'Tuan was conceived at last'.[4]

'When the Queen saw the pure salmon
She desired him.
So that was assigned to her, a good course,
And she ate it all by herself;
The very noble queen became pregnant,
And thence Tuan was conceived.'

Thus Tuan the fish, loved and devoured, became Tuan the pure and royal human child, reclaimed in a female womb, doubling as the world's ocean. A proto-Tuan, travelling from the deep underwater world, could then arrive in the 'land-world, equipped with the knowledge of submarine wholeness, and fit to recreate a new human order, with his purified bloodstream'.

In zoological fact, the annual migration of salmon from the west to north coasts of Ireland, passes Grianan Ailech, the queen's northerly home, where, in eating the fish, she touches it with her tongue. And the word 'tongue' is cognate with 'fish' in several Indo-European languages.

Thus in Achill island, County Mayo, the sole fish is called *teanga cait*, literally 'tongue chaff', to emphasise the need for verbal realisation of mythic ideas, since a deity is born when 'word of mouth' and delivery from the womb are twinned together.

In Nature, the male salmon comes from the sea in late winter, in good, fat condition. Leaping waterfalls up to three metres high, he reaches the spawning place, usually a gravelly reach in the upper stream, chosen by the waiting female. With her tail she excavates a hollow, 30 cm deep, and a metre long, into which the male drops his milt onto her roe. After the female covers the gully, they both then repeat the process.[5]

So river and landscape act as a surrogate mother to this annual event. 'Tuan's' 'wholeness', arises when both his 'fishy parents' couple with Eriu, Ireland's land goddess, every year. The whirlpool of the natural cycle provides the shrine, where gods and humanity can rediscover their sacred worth, under the sign of the fish.

LADY OF THE PLANTS –
CHRISTIANITY'S LINKS WITH THE
ORGANIC WORLD

So many contradictory attitudes have been laid on Britain's flora, that it is almost impossible to discern a distinctive Goddess assignation among the tangled attributions. However, there are numerous examples with traces of Classical inheritance; discernible for example, in *Labrum Veneris*, 'Lip of Venus', the teasel's gripping seed head, and 'Basin of Venus', for the same plant's water-retaining concave leaves.

That deity also features as 'Venus in her Car', a Somerset name for the Water Betony plant, and she rides in 'Venus' chariot drawn by doves', an alternative name for the blue flowering Monkshood.

The corn bellflower is also known as 'Venus' looking glass', due to the shiny texture of its flat seeds, while the Pennywort doubles as 'Our Lady's navel', a name derived from the Roman *Umbilicus Veneris*. Finally, we claim a share of the Grass of Parnassus, Parnassus being where the Muses once sported.

No fewer than 132 British wild plants are attributed either to 'Ladies' or to 'The Lady', while a further 19 are called 'Queen'. To what extent these females make supernatural claims is unclear, but the Anglo-Saxon habit of referring to their deity Freyja, as 'The Lady',[1] should be remembered, but not confused with the Catholic 'Our Lady,' (13 instances) alias 'The Virgin Mary', (who is linked to another 12 plants).

Lady's Bedstraw, Galium verum. *One of many wild plants to which The Lady's name is attached.*

Christianity has clearly led to the suppression or change of previous floral nomenclature. Thus no fewer than 87 species are now handed to the 'Deil' or 'Devil', in order to counteract their positive pre-Christian associations; and with a similar unease, 54 species are now ascribed to the 'Fairies', while Witches are connected to 20 herbs.

Regardless of some attempts to Christianise many plant names, the seasonal upsurge of blossoming vegetation, sweeping across the land every springtime, was still an unstoppable pagan event. In addition, the annual April arrival of the cuckoo (who features in 72 plant names), served to bring sky and underworld powers together, in a collaboration that produced the splendid display on the earth's surface – an abundance that included a surge in animal and human vitality.

Thus 'Easter' has 10 local plant names, 'Jack' rises in 40 vegetables, 'Kiss' occurs in 26 names, while 'Love' features in another 26, and the month of May is registered by a further 35 species. Milkmaids claim 15 showings, yet 'baby' is seen among the flowers a modest 12 times.

Thus, in her reproductive capacity, 'The Lady' rides again, emerging from whatever stable she acknowledges; and as a 'Mother', she rates 210 mentions. The less fortunate are not forgotten. Thus 17 'Poor Men' feature in the British plant name list, along with eight 'Beggars'. There are also 29 'grandmothers' or 'Grannies', and 30 'Grandfather' or 'Old Man' labels. So the entire human lifespan is respectfully covered, including 'The Dead', mentioned in 31 instances, and 'Blood' or 'Bloody' which also occurs 31 times.

What also emerges from scrutinising these British local plant names is people's willingness to include both domesticated and wild creatures in their choice of names.

This is in keeping with the all-inclusive life song of the great Goddess, and with her own marked ability to manifest in animal form. Thus pig and sow (a Freyja nickname[2]) swine and hog, together carry supernatural connections into the plant kingdom a combined 76 times. Dog claims 77 plants; cow 57; the cat, ranging as it does between worlds, 46; horse 39; cock 37; bull 36; sheep 12; while the crow, as a battlefield scavenger, has 47 plant connections. Various other birds claim 26; the hare has 10, and the rabbit only six.

Of wild creatures, the snake is associated with 59 different plants, and another 14 are specifically given to adder/viper. Dragon has seven types of flowers, while robin, with his uncanny Robin Hood links, has acquired 25 plants.

A widespread fondness for discovering inter-species connections that cut across 'normal' animal-vegetable-mineral boundaries emerges from this brief study.[3] The links provide evidence for an instinctively felt unity, divinely prepared, and daily experienced, of an organically wholesome world.

GODDESS AS SNAKE

In our Christian country, ruled over by a monarch who is the 'defender of the faith',[1] people read the Bible's Book of Genesis, in which it states that Eve, (Hebrew *Hawwah*), is the 'mother of all living'. It is she, the Bible implies, who is both the mother of humanity, *and* of all plants and animals now alive[2], *or* who have ever existed, in this world. Surely the breadth of her achievement entitles us to regard Eve as a goddess, or even as the Great Goddess, the primal Creator?

In this, her labour, she is rightly associated with the serpent, for that creature provides a symbol of energy, an energy dressed in its myriad disguises. Snake and woman, acting together,[3] achieved the colossal task of filling the entire planet with life.

But according to the biblical account, this could happen only after Eve had followed the serpent's encouraging advice, and had enjoyed the fruit of forbidden knowledge.[4] Then, working as a pair, there was no stopping them, despite warnings from their patriarchal opponents, represented by Adam and a transcendental Jehovah, who jointly produced the doctrine of Original Sin as the Eve-Snake partnership's penalty.

In fact, after its brief entry into the Garden of Eden, the serpent did much more than slither down to Hell; for in addition he found many congenial refuges in an England,[5] pervaded by the belief in serpents as enormous 'worms'. These reptiles have given their names to major topographical features, such as: Worm's Head, and The Great Orme, both in Wales, and Wormigill and Wormelow in England. These creatures operated in and as England; while, a winged 'dragon', derived from the red snake, the badge of their patron, St David, became the flying serpent, adopted as the national emblem of Wales.

In Northumberland, the Lambton Worm was fished out of the river Wear, by the lord of Lambton Manor, and *on a Sunday*, when he should have been in church. Some people therefore believed that this worm was the Devil incarnate. Having outgrown the well into which the Lord had flung his catch, the monster then encircled the nearby Worm Hill three times, before desolating the area. Eventually, the milk of nine cows was provided for it, in a trough, rendering the beast peaceful. When further attempts were made to kill the worm, its body, cut in two, simply reunited.[6]

In Scotland, the worm of Linton in Roxburghshire has a den named Worm's Hole in a local hill, and a carved image in Linton

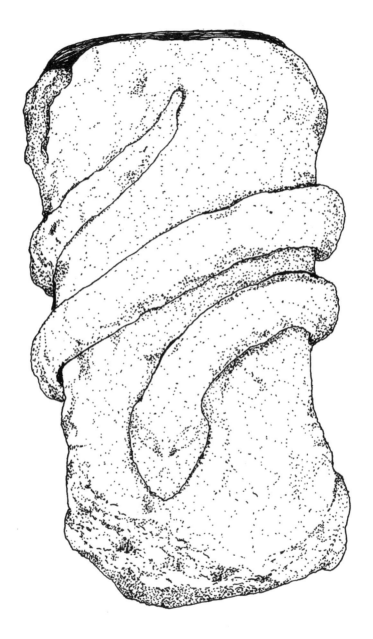

A Roman 'snake altar' from the Severn valley. It is now in Stroud museum.

church. That worm was killed by a man who attached a pitch-soaked flaming wheel to the tip of his lance, and rammed it down the monster's throat.[7]

On a country-wide scale, after the Meister Stour Worm had been killed in Orkney its body became Iceland, and its teeth turned into The Faroe Isles.[8]

Much smaller, in 1614, near Horsham in Sussex, a strange and monstrous serpent or dragon left a track of glutinous, slimy, ill-smelling matter. This serpent was reputed to be more than nine feet (2.7m) long, with red scales on its belly, and black scales on its back.

Another Sussex worm, the *nicor* (Old English for water monster) reputedly lived in a 20-ft (6m) wide bottomless pool named Knucker Hole at Lymington. In these examples, the recurring theme brings water and the subterranean together with rumours of buried treasure. This underlies a superimposed Christian layer of fearful serpent rejection.[9] The emphasis on serpent size shown in these folk tales fits their topographic engendering roles. As primal earth-builders they *are* the living essence of the landscape. With such elemental forces at play, one cannot attribute the fear that they produced entirely to Christian influence.

By contrast a positive view of snakes, allied to the maternal procreative act, was normal in Graeco-Roman culture. Thus amongst the rediscovered Backworth (Norfolk) hoard of Romano-British objects,[10] a golden snake image was found, inscribed to the *Deae Matres*, 'Mother Goddesses'. Likewise in the Snettisham hoard in Tyne and Wear,[11] snake imagery appeared among numerous intaglio brooches, bearing the image of the Roman corn goddess Ceres.

Moreover, in Roman Britain snakes were kept as pets. They were associated with the *Genius paterfamilias*, and featured in the typical domestic *Lares* shrine, to express the extended family's well-being.[12] As symbols of rebirth, good health, and material prosperity, a pair of intertwined serpents also formed Mercury's *caduceus*, or ceremonial wand, his symbol of prosperity.

Snakes also accompanied the god of Health, Asclepios, and embodied the cures attributed to him. Consequently, serpent-headed gold bangles and intertwined snake brooches, feature in both Roman and Celtic metalwork.

In fact, the Romans did not *introduce* snake worship to Britain; rather they emphasised an already well-established tradition, present here since the Stone Age. For example, the twin serpentine stone avenues that converge on the henge monument at Avebury, Wiltshire, were constructed in the middle of the 3rd millennium B.C., and demonstrate the snake's inclusive role in providing a living vehicle for mass human ceremonial events.[13]

Similarly, a serpentine avenue revealed at Rotherwas, near Hereford, during roadworks in 2007 was a zigzagging mosaic pathway, thought to be at least 75 metres long. According to the archaeologists, this snake was 'deliberately sculpted' to undulate up and down and had a surface of fire-cracked cobbles, interspersed with white quartz pebbles. The cobbles had been heated in a hearth and then dropped into water, in an abrupt elemental collision.

Aligned north to south, this 'snake' was predictably termed a 'ribbon', and therefore lifeless, by local archaeologists. The description 'snake' epitomises vitality, and so might attract pagans, or enthusiasts such as Dr Maria Vazquez-Hoys of Madrid University, who describes the Rotherwas discovery as 'the most important religious site known to me, for the study of serpent cults'.[14]

The Rotherwas monumental reptile directly addresses the river Wye, and, as its lower layer shows, it was rebuilt at least once, with a lifespan running between the Late Neolithic into Early Bronze Age, c. 2000 B.C. Potsherds found in the immediate vicinity extend the period of use back to 3200 B.C. This serpent has been compared to the Serpent Mound, in Ohio, USA.[15]

Sadly, later in 2007, Hereford City Council ignored public outcry and completely buried their inconvenient 'serpentine Ribbon' beneath a new relief road, covering the feature in concrete 'for posterity's sake'.

The fiery treatment of the Rotherwas stones may relate to the cult of the fire goddess, Brigit or Bride. In Scotland Brigit was believed to emerge annually from under a hill, and in serpentine form, on her February quarter-day festival, an act paralleled by the snake casting off one skin before revealing a replacement.[16]

Another snake-like Scottish earthwork, the Loch Nell Serpent mound, lies four miles south of Oban, Argyllshire. The serpent was described by Constance Cumming in 1883, as rising conspicuously from a flat grassy plain. This 'totally artificial structure is about 5.18m–6m (17–20 ft) high, and 91m (300 ft) in length, was composed of regularly and symmetrically placed stones, set like the vertebrae of some huge animal, with smaller stones suggestive of ribs', and with a view across the loch to the triple peaks of Ben Cruachan.[17] On the serpent's head was a circle of stones, within which three large stones formed a megalithic chamber, excavated by J.S. Phene in 1871. The monument was built c. 2000 B.C.

The goddess-as-snake combination may be seen again, on a smaller scale, in many of Britain's prehistoric rock carvings, where, from the concentric rings around the central cup of a feminine 'container-body', a serpentine line, acting as a water-filled gulley, enters into *and* emerges from this maternal figure.

In 19th Century Wales, as Trevelyan reported, a snake would be seen slithering away from a farmhouse upon the death of the farmer or his wife.[18]

In John Ruskin's study of the Greek snake deity Medusa and her link to the goddess Athena, he starts by accepting the living world depicted by the English Romantic poets Shelley, in *Ode to the West Wind*,[19] and John Keats in *Lamia*,[20] which tells of metamorphosis between a woman and a snake. Both poems describe a world of differences, united by a serpentine life force that can be delivered by the wind, or by human breath, serving either as an avenue for poetic utterance, or for music, set to divine words. Such messages are carried by, and depend on, snaky, invisible waves of air; so too, is every word of myth narrative. Or as Wordsworth put it, there is 'a motion and a spirit, that impels all thinking things, all objects of all thought, and rolls through all things'.[21]

Similarly, the placing of the Gorgon's snake-surrounded head upon Athena's breast-plate was emblematic of a snake and goddess, defining a spirit-matter fusion that they together shared and embodied, and was specifically displayed by the *mortal* Gorgon, with her snake-framed head.

Consequently, as an antidote to mechanised 19th Century England, Ruskin turned to, and venerated, 'the everlasting lamp of Athena', whose shape-changing snaky powers balanced creation and dissolution.[22] Meanwhile, in womankind he had discovered the human display of volatile changeability, comparable to, and echoing, the idealised dynamism that he saw in Athena.

Yet, as with a British cup and ring prehistoric rock engraving,[23] and its penetrating phallic channel, Ruskin's feminised world depended on sexual differentiation. In order to revivify the feminine, he understood that communion must be preceded by separation, prior to sexual intercourse, when 'she' ingathers and harvests her opposite. So, the same old snake, with a vertical salute of recognition, comes home, by re-entering and reanimating the Goddess's world.

HOLY COWS

Every Saturday morning in 1954, I drove Mr Williamson's herd of thirty dairy cows from the milking byre, to their field behind the Baptist's chapel, before returning to the farm to clean out the splattered byre. I enjoyed this weekly cow-walk down Green End Street, watching from behind the easy rolling motion of their pelvises, as they sauntered along, indifferent to the occasional van wanting to get past.

In those far off times each animal was known by name (such wild flowers as 'daisy' or 'buttercup' were commonly chosen), rather than by a plastic number tag. In 1954, cows were more than mere milk machines.

Yet I was unaware that these beasts' ancestors had walked all the way from India, via the Phoenician shore of the Mediterranean, bringing with them a trail of myths, which were still being retold across two continents. By chance, I was steering a herd of goddesses, or at least the great-grand-daughters of divinities, who deserved to be held in high regard. But having recently emerged from childhood, I was in no mood to hear, far less believe, more fairy stories, and preferred to think of the animals in my charge as plain cows, of a black and white Friesian breed, from the near continent.

Subsequently, I learned that in Britain and Ireland, *magic* cows were usually pure white, with red ears.[1] A herd of that colour, surviving at Chillingham, Northumberland, is rumoured to be descended from a white herd, introduced by the Romans to meet their ritual needs and serve as sacrificial offerings. (Alternatively, these truly wild beasts may be descended from aurochs, and were enclosed in a walled park only in the 13th Century.)[2]

But why are the supernatural cows of British and Irish folklore so often described as white? White, the colour of milk and snow, is associated with purity and innocence. But just as white is produced by an amalgam of all the colours of the rainbow, plaited together, so white's 'purity' is inseparable from wholeness, and completeness; made from everything; in a divine paradox it depends on the obliteration of every other colour. Further ambivalence stems from white as the colour of shrouds, producing a life-death partnership.

In addition, the generosity of cows is so appealing. They give milk, cheese, yoghurt, and butter, purified into clear ghee, used in sacrificial fires. In addition they provide meat, hide, urine (used medicinally in India), and act as draught animals in ploughing. They also provide manure, an invaluable fertiliser, and as an ingredient in the plaster of house walls. Cows offer all these gifts, yet they demand so little – only grass and water.

In modern India, cows are deeply appreciated. In their old age they can chose between over 3,000 charitably supported '*Gaushalas*' – old cow retirement homes.[3]

They are also worshipped as Prithivi, a bovine manifestation of Mother Earth, and as Kamadhenu, the archetypal female cow of plenty, often sculpted with a human female head and breasts. She is regarded as the Mother of all cattle.[4]

In Ancient Egypt, the sky-goddess Hathor was seen as a celestial cow, who sheltered the sun god at night, and gave birth to him again at dawn. In addition, she was credited with creating the entire word.

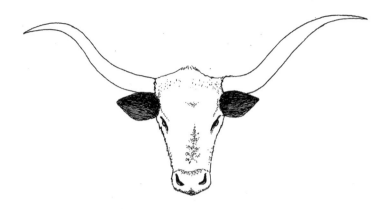

A long-horned cow's head.

Sculptors portrayed her with a horned woman's head, and with a cow's ears.

Hathor was a goddess of joy and love. She was considered the mistress of merriment, including dance, music and song, leaping, and garland making. She held a sacred rattle called a sistrum, which, when played by her could drive away evil spirits. Her chief temple was at Dendera. Hathor suckled every Pharaoh, and cherished the dead. As Queen of the West, she welcomed the dead into the Other World.[5]

As the Lady of Byblos, her influence spread to Phoenicia, where as the princess Europa she took a ride upon the back of the god Zeus, in his bull form, and was transported during his sea voyage to Crete. She became Crete's first Queen, as described by Ovid. She subsequently gave her name to the entire European continent, including the British Isles.[6]

Just as pastoralists drove their herds to Europe's Westernmost Isles, their beliefs came with them, for the well-being of the animals depended on a continuing connection with their divine prototypes – even if the expression 'Holy Cow' has now sunk to a slang term, bandied about by American baseball players.[7]

Christianity also adopted the sacred White Cow image as a working reality. Thus Ireland's St Kevin, born in 398, was fed every morning and evening by a mysterious white cow, who visited his parents' home. He went on to become the first abbot of the Glendalough monastery.[8] Likewise, in her infancy, the pagan goddess, Brigit, was fed by a fairy cow.

Irish, Welsh and English folklore records numerous instances of the White Cow's numinous influence. On Wales's highest central mountain she is recognised, along with her calf, in the pair of massive quartz stones, named in her honour.[9] Irish sources describe the yearning to encounter the animal's elusive mountain-top presence.[10] This reminds us of Ninhursag, an early Sumerian cow deity, whose name means 'the Lady of the Mountain'. She presided over a dairy-temple, and nourished kings with her milk.[11] Similarly, at the Mitchell's Fold stone circle in Shropshire, a white cow would appear in times of famine, and willingly fill vessels of every size with her milk, without charge.[12]

English feelings for the White Cow have been supported by the mythic Norse figure of Audumla, wet nurse of the giants. Audumla fed Ymir, the original androgynous human, with four rivers of milk,[13] enough to satisfy the most demanding of contemporary dairymen's production targets. Dear cows, you carry in your bones qualities and values accumulated during thousands of years, and from around the world, as you wander down Green End Street, on your way to producing tomorrow's milk.

BEES: POETRY, MAGIC AND REGENERATION

Of the billions of types of insects in the world, the bee is the most closely connected to the Goddess, a link maintained since Stone Age times, and still reverberating in British folklore.

There, the Queen Bee, who gives birth to all her subordinate workers, is regarded as the epitome of creation. In England, some bee-keepers talk to their bees, and treat them as members of the human family, and inform them of important family events. For example, a bride should personally tell the bees of her forthcoming wedding, and be sure to leave them a slice of her wedding cake. The birth of

a child, and a death in the family would also be reported to them, with wine and food from the funeral given to the swarm. Failure to do so might cause the swarm to die, depart, or lead to a decline in honey production.

Bumble bees are still regarded as promotors of human health and prosperity, and as reliable weather predictors. Because of their vital role in the pollination of plants, (a task now seriously threatened by pesticides), they are rightly seen as guardians of the natural world.

Because they retain a strong connection to a pre-money culture, a swarm of bees should be given away, or bartered, but not sold for cash. In return, they continue to bring with them associations with magic, love, and industriousness; and through honey's incorporation into mead bees are linked to creativity, poetry, eloquence and song.

Mellona, the Roman goddess of bees, derives from Melissa, the Greek nymph who discovered and taught the uses of honey. Melissa also cured the sufferings of women in labour, and was associated with the moon goddess Artemis. In her old age, Melissa became a priestess of the Goddess Demeter, whose secrets she declined to reveal, whereupon she was torn to pieces by women. Demeter then caused bees to be born from Melissa's dead body. Consequently souls became symbolised by bees, and Melissa drew these ethereal spirits down to be born.

Melissa was also connected to the power of periodic regeneration. Her bees were able to see into the future, and were regarded by the Greeks as 'birds of the Muses', and as messengers from divine realms. Therefore, in English folklore, if a bee flies over the face of a sleeping child, that child will enjoy good fortune.

THE WHITE LADY IN BRITAIN

What happens to female deities after they are ousted, forbidden, and redefined as evil, by an incoming male monotheism? Obliged to vanish from the new norm, they sometimes reappear, either as ghosts, or as fictional characters, sustained obliquely in oral folklore; or they arise like mushrooms, as subconscious perturbations, through the unstable surface crust of rejection and denial.

So for example, in Christian Britain, the deposed Great Goddess can be glimpsed as The White Lady, and Welsh *Y Ladi Wen*,[1] alias *Dynes Mewn Gwyn*, the 'Woman in White'. Together these ghostly shadows transmit intermittent rays towards lost glory, and incorporate into their widespread manifestations the vestigial traits of their previous functions.

It could be argued that these White Lady phenomena reflect nothing more than the white burial shroud of ordinary funerals. If this is so, one is struck by the complete lack of any 'White Gentleman' equivalents. Perhaps *they* are not needed, since maleness now supervises divine and current affairs. By contrast, The White Lady tends to behave as a self-defined outcast. Her appearances are often tentative – compelled to come back, but often too shy for a face-to-face encounter with modern humanity.[2]

Thus among five great pine trees, close to the Wimborne to Cranbourne road in Dorset, she is seen with a white hood over her face, pacing to and fro. But if approached, she dashes through a hedge, after which the crashing sounds of a coach and horses are heard, gradually fading into the distance.[3] (It will be recalled that

A Roman Venus figurine, as the basis of a White Goddess image.

Nerthus, the female Earth Goddess, of Anglo-Saxon pagan times, usually did her rounds in a horse-drawn wagon.)[4]

By contrast the White Lady of Longnor in Shropshire lived in an allegedly bottomless pool, but frequently emerged to dance on the village green.[5] An association with water is also demonstrated by a Derbyshire White Lady, who haunted the confluence of the Dove and Hodge rivers. At Hallowe'en she manifested at that spot as a beautiful maiden, with long golden hair. Her arms were usually thrown around a white doe. Pre-Christian Hallowe'en corresponds closely with *Samhain*, the feast marking the start of the pagan year, when the Triple Goddess typically assumed a maidenly form.[6]

Her maternal qualities are evident at Blenkinsopp Castle in Northumberland. There the young son of the place found the White Lady sitting on the end of his bed one night. Weeping and wringing her hands, she kissed the boy, and urged him to go with her, to find a large box, filled with gold, buried in the castle vault. She could not rest until it had been relocated and passed on. But as she lifted the boy up, he screamed, which frightened her away. The same thing happened on the two following nights; after that the boy slept in his parents' room, and the White Lady tried no more.[7] What was the treasure, perhaps a symbol of the recurring underground wealth of annually germinating harvest, the agrarian basis of the mother goddess mythology, rooted in the Neolithic, and here rejected by the next generation of landowners?[8]

Gold features again, at Ogmore Castle, near Bridgend. There a White Lady showed a man her golden treasure, hidden under the tower floor, and said that he could have half of it. He took half, but later, overcome with greed, he returned and made off with the rest of the hoard. Whereupon, the White Lady reappeared, with her fingers changed to claws, and all-but tore him to pieces. He returned home, but he rapidly wasted away from shame,[9] and soon died. The concerns of fair sharing, moderation, and the honest keeping of agreements are messages that were largely ignored in the depletion of the mineral resources of a now exhausted South Wales.

That her flowers were of high value is suggested by a White Lady who appeared to a shepherd near Llantrisant.[10] He saw this white-robed girl scattering flowers, and after she smiled sweetly at him, he picked the blooms up and took them home, to appreciate their beauty. By the morning they had turned into gold coins.

White Ladies also attended to their original function as divine receivers of the dead. So near Llanerfyl, at the location where several men had been killed while building a new road, she appeared at midnight, in a pure white, trailing garment, and performed a commemorative dance.[11] Likewise, at Wellow, Somerset, she performed a mourning role, to the Lords of Hungerford and their families, while haunting the nearby St Julian's well, in a cottage garden.[12]

Sometimes the White Lady rode on a white horse, as at Corfe, in Dorset, where her night-time gallops were famous. There 'She shone like a dew-drop, and sounded like liddle bells all a chime', according to people in the Blackdown Hills, Dorset.[13]

The link with fire is another aspect of her persona, and one that recalls her former solar connections. As at Llangwym, in

Denbighshire, North Wales, she disappears into 'a great ball of fire' when a youth casts doubt on her existence;[14] while across the town of Burford in the Cotswolds, a Lady Tanfield rode over the rooftops in a fiery chariot, foretelling misfortune, until, in the 18th Century, seven clergymen pushed her into a bottle, corked it, and threw it into the river, from where she will arise in times of drought, to continue her rooftop journeys.[15]

Many White Lady sightings are less spectacular, and lack historical connections. For example, in Warwickshire's Ragley Hall, a White Lady quietly sits on a stile, while drinking water from the river Arrow. The stile is a liminal feature, made for accessing between different 'fields' and alternative modes of being.[16]

In Wales, The Lady's 'Wen' title is a shortened form of *gwen*, Welsh for 'white' and 'holy'. *Gwen* is also found in the name Gwenhwyfar, (Guinevere), 'white phantom', Arthur's wife.[17]

Y Ladi Wen's accumulated sacredness, now rarely recognised, may explain why she was often seen at night, wringing her hands in great distress. She carried such a heavy load of references. On one occasion, a man asked her what he could do to help. She answered that if only he would hold her tightly by both hands until she told him to stop, her troubles would be over. He grasped her, until a barking dog distracted him, and he lost his grip, whereupon she screamed, and said: 'I shall be bound for another seven years!' and then she vanished.

Like the after image of an extinguished bright light *Y Ladi Wen* continued to sit at boundaries, such as cross roads, fords, and in church yards, unwilling to admit that her task of reconnection was over. By 1900 A.D., her last vestiges were left to children, who knew her as a wraith, a bogie, who, they were told, could still arise to punish their misbehaviour. So they took away even this threat, by singing her into nonsense:

'*Y Ladi Wen* on top of a tree is sawing an umbrella shaft. It's one o'clock. It's two o'clock. It's time for the pigs to have dinner!'[18]

In this rhyme she turns into pigs' swill. Yet from that bin, she could leap out as *Creiddylad*, beautiful daughter of the God Lludd Silverhand, 'the maiden of most majesty there ever was in the island of Britain'.[19] Her name combines Welsh *craidd*, 'centre, heart, essence' with *Y Ladi*,

'The Lady'.[20] Upon her strength, young women continued to draw, during the May Day dances, around the Merry Maiden stone circles, erected in the prehistoric Bronze Age.

Hence another children's rhyme cries out: 'There's Mother, coming over the white stone', offering a *maternal Ladi Wen* – menhir link, with both whiteness and fecundity drawn from the rocks.[21]

The emphasis on whiteness in images of the divine feminine, is also a feature of the innumerable white clay Venus figurines, found throughout Roman Britain.[22] Nor was the special quality, associated with that colour, confined to super*human* imagery.

Thus reverence accorded to white horses, as equine representations of the deity, can be matched by a corresponding regard for white cattle, as seen in the 1.67 metre (5 ft 6 in) tall White Cow of dazzling quartz, who sits with her quartz calf, upon Pumlumon, Wales' central mountain.[23] Together they epitomise and concentrate the sanctity accorded to all cattle. Similarly, on May Day morning in Ireland, people praised the White Cow.

In Shropshire, one of the stones in the Mitchell's Fold prehistoric stone circle is known as The White Cow. There, during a famine, a magic cow had given milk to all and sundry, so long as no one took more than a pail full; but when this limit was broken, the animal turned to stone.[24] The theme of measured fairness is embedded in White Lady lore, inherited from the Neolithic era, characterised by a collective responsibility, that she tacitly advocates.

Meanwhile, the Rough Wood Nature Reserve, in the industrial Black Country, was haunted by a White Lady, who came soaking wet, from the pool where she was drowned. She was the spirit of Pauline Kelly, who, along with her daughter Evelyn disappeared in the mid-19th Century. The local community has a tradition of wearing white dresses at Hallowe'en, and chanting the mocking rhyme: 'White Lady, White Lady, I'm the one who killed your baby'. This sardonic performance could also be interpreted as their communal *merging* with the White Lady persona, while nevertheless killing off her child at the start of the pre-Christian New Year.

The sacking of resources of coal, iron ore and limestone during the Industrial Revolution, as witnessed in the Black Country, matches their murderous claim. Yet their performance is specifically sited near the Black Country's only nature reserve and there they re-enact the drama as *participants*, rather than mere observers. The White Lady's return, and her fate, along with our own, still hangs in the balance.[25]

> When the Lady knocked;
> Then we saw the outer hall
> Grow lighter
> And she was standing there
>
> …We have seen her
> The world over
> ….And the point of the spectrum
> Where all lights become one
> Is white and white is not no-colour,
> As we were told as children,
> But all-colour;
> Where the flames mingle
> And the wings meet, when we gain
> The arc of perfection,
> We are satisfied, we are happy,
> We begin again.

From *Tribute to the Angels*, by H.D. (Hilda Doolittle)[26]

THE DIVINE SUN

To compare the Blessed Virgin Mary to the elemental all-embracing Air, encountered with each breath we take, is a view largely confined to poets and mystics. By contrast, to regard the sun, which is our chief source of heat and light, as a goddess, has long been normal on every one of the world's continents. In Asia, Amaterasu, 'great shining heaven' is the head of the Japanese Shinto pantheon, and features on

the Japanese flag.[1] India has Aditi as keeper of the light that ensures life and consciousness.[2] Africa, Australia and North America each have a number of female solar deities.

In Europe, the Lithuanians and Latvians continue to worship Saule[3], their goddess of the summer equinox.[4] Her counterparts are Paivatar in Finland[5], Beiwe in Lapland[6], and Xatel-Ekwa in Hungary, where she rides horses through the sky.[7] To the Basques, the equivalent sun goddess is Ekhi.[7] In all Germanic myths, sun deities were female, whether known as Sunna[8] or, (among the Norse), as Sol, who rode the skies in her chariot.[9] Norse and Germanic traditions took root in England from A.D. 400 onwards. Here, not until c. 1600, did the sun turn masculine, due to Renaissance Classical influence.

In Celtic Ireland and Scotland, female sun deities remain firmly installed in those languages. In short, the male-female division of the elements sometimes proposed, which nominates Air and Fire as preserves of male deities, with the goddesses restricted to Earth and Water, fits neither the inherited traditions of the British Isles, nor the observable interplaying unity of the elements.

In Gaelic-speaking Ireland and the Scottish Isles, the Goddess is said to bring all the elements together, and combines them into her cosmos-wide working metabolism. She thereby creates, and is synonymous with, the living totality, and all the life-forms contained within its scope; for them, air, heat, and light, are indispensable ingredients. As if to emphasise this earth-sky interdependence, the Irish word for 'sun', is *grian*, (and Grian, is one of several Irish solar deities); yet *grian* also means 'the gravel in the bed of a river'. Her influence runs vertically, from the sun, through water, to bed rock.[10]

Grian's sister Aine, (whose name is interpreted by O'Rahilly to mean 'she who travels regularly'[11]), reflects a pre-Copernican, earth-centred kind of cosmos, while bringing the seasonal dynamic into play. Together Grian and Aine, integrate their vertical and horizontal axes, as described by the sun, from bedrock to apogee, and from midwinter to the summer solstice, when Aine is annually witnessed and celebrated, as she arrives over her Cnoc Aine sacred hill, in County Limerick.

A solar deity among the stars.
The sun has been worshipped as female on every continent.

To modern Goddess hunters, the range of their deity's web suggests an opportunity to discard the cult of modern individualism, the self-centred ME, by merging its sharply defined boundary into her own expansive, volatile, nature. She offers a life-long ride on the merry-go-round of her huge reality. So placed on the elemental Goddess, the rider may adopt some of her generosity and breadth.

The process of synthesis with the Goddess is rendered a far less formidable challenge than might be supposed because her movements are not regarded as abstractions, but have positive qualities and emotions associated with them. For example, the goddess Aine's sunny rotation, the Irish 'Dessel', also means 'lucky, favourable, and propitious', and brings connotations of 'seemliness, comeliness and beauty'. Similarly, the Irish name Aine also serves as the Irish word for 'delight, joy, pleasure, agility, swiftness, play, sport, music, experience, truth, brilliance, wit and radiance', providing plenty of incentive to those who contemplate the possibility of joining her solar troop.[12]

Further encouragement to the individual, seeking integration with the Goddess, can be found in the frequency with which she is willing to deliver small scale, or domestic versions, of her far-flung self. So the mighty Goddess of harvest hangs over the cottage table in the form of the corn dolly, alias the Cailleach or great hag[13], while the sun's awesome fire ball reappears as the flame within Brigit's household lamp.[14]

Brigit, another solar deity, is 'the great or sublime one', whose name derives from the root *brig*, 'power, strength, vigour, force, efficiency, substance, essence, and meaning'.[15] She is both a woman and 'a flame, golden, delightful', venerated in particular on St Brigit's Eve, and day, the February 1st mid-point between the winter solstice and March equinox, when the sun, growing in strength, brought hope of winter's end. During that festival, named after her, she re-entered each house, in the shape of a doll 'child' that was placed in a specially prepared cradle.[16] So, even the mightiest embodiment of the Goddess can, and annually did, turn up on the doorstep, hoping for shelter in every home and heart. Her very existence still depends on a two-way process, involving every household, and each of its members, who one by one contribute to her universal love affair. Her macrocosm is nourished by acts of devotion coming from these microcosms.

II.
The Dancing Goddess

AN ARCHAEO-MYTHIC GRAMMAR

Most games have rules. Listed below are some rules, known since prehistoric times; but as in any playground, different games can be played simultaneously, so you may anticipate confusion.

Attempts to trek (backwards?) into an understanding of the prehistoric world have been rendered difficult for educated Westerners, partly due to the long-term influence of the Greek philosopher Socrates, as disseminated by his pupil Plato. Plato's Athens Academy, founded 387 B.C., downgraded myth-centred experience and replaced it with Socratic *logos* that emphasised linear, rational, objective reasoning.[1] This preference was turned into dogma by the 17th Century cult of Reason[2] and was further entrenched by modern scientific instrumentalism and abstract theory seeking.[3] This approach has been adopted by modern academia worldwide (including in departments of archaeology), to become the dominant methodology, within and beyond academic confines.

As a consequence most recent archaeological studies of pre-Socratic myth-based societies have been hampered from the outset by an anti-mythic outlook. Yet in the last three decades some archaeologists have come to see the mismatch between the avowed aim and the chosen method of prehistoric research as an absurd incongruity that blocks, rather than illuminates, an understanding of the ancient world.

The sense of frustration is expressed in Ian Russell's *Images of the Past* (2006). He writes of the current 'crisis' within archaeology and 'the fundamentally problematic basis of the discipline'.[4]

Similarly, proponents of the technologically obsessed 'New Archaeology' of the 1960s and '70s admit in retrospect that it produced only a sterile, mechanistic functionalism – a mirror image of the practitioners' own positivistic limitations, rather than insight into the prehistoric mind.[5] Yet despite this belated recognition, in the British Isles, archaeology remains stuck in an almost paralysed state.

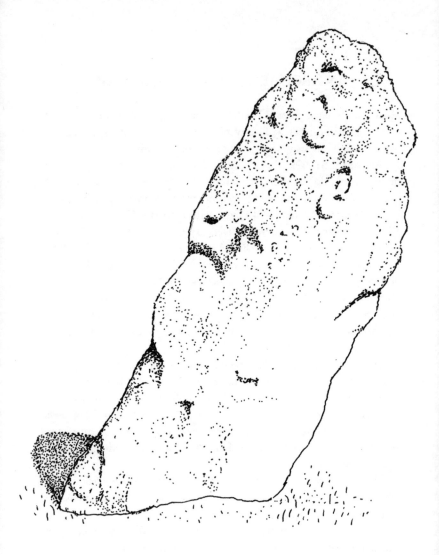

The Stonehenge Heel Stone is 4.9 metres high. In supernatural terms this megalith aligns with both the midwinter sunset and midsummer sunrise, and so combines contrasting happy and sad expressions, to match the solar deity's changing moods, both underground and overground. It was probably the first megalith to be erected on the site.

As Julian Thomas, Professor of Neolithic Archaeology at Manchester University, rightly laments, our archaeology measures 'a world of bare material things... an array of dead substance... quite separate from the realm of meaning and value'. Thus by its insistence on logos-derived 'objectivity', archaeology has been 'complicit in a process of disenchantment'.[6]

British Archaeologists find it hard to acknowledge *the pre-modern*, mythically experienced life-world, since myth fails to register on their instruments. Meanwhile, in general usage, myth has acquired a very bad name. The very word 'myth' is currently defined as 'an untrue or discredited story or belief; a misconception; a misrepresentation of the truth', or 'a belief without foundation'; or 'a purely fictitious narrative, usually involving supernatural persons, actions or events'.[7]

If the 'supernatural' is now meaningless to the secular modern researcher, a corresponding disparagement of myth on a popular level in Britain stems partly from more than a millennium of intolerant Christian monotheism. In addition, the Protestant Reformation introduced an aversion to any form of sacred imagery in England. (Today, with or without 'idols', British Christians fail to see that Christ's annually repeated cycle of birth, maturity, sacrificial death and resurrection is a typically mythic construct.)

Monotheism, scientific reductionism, and preference for an emotionally repressed 'stiff upper lip' style of delivery have combined to deter most UK archaeologists from stepping into mythic realms that would give prehistoric meaning and *feeling* to their dry tabulations.

The solution to this impasse has been demonstrated by the California-based *Institute of Archaeo-mythology*, in reconnecting modern archaeological research to the myth-based structure of Antiquity. The Institute has led the way in evolving an interdisciplinary approach, and a pluralistic, more appropriate methodology, that has brought Antiquity back to life and has provided an encouraging paradigm for the disheartened archaeological mainstream.

As one among many associates of The Institute who has benefited from its attempts to achieve a reunion of archaeological evidence and myth, I offer the following brief summary of some aspects of the underlying 'grammar' that myth-based cultures may share. These findings are derived from (i) my country wanderings, encountering many numinous natural features; (ii) some archaeological labouring; (iii) numerous visits to prehistoric sites, temples and museums; (iv) many decades spent in library research; (v) participation in folk festivals; (vi) frequent involvement in community arts projects; and (vii) attendance at some of the Institute's conferences, organised by Joan Marler.

A Pocket Guide to Myth

1. The primary words and narratives conveyed by myth are regarded as being uttered and dramatically performed by the deity or deities, when in the process of creating the entire cosmos, during a *cosmogonic* event.

2. Consequently the universe is seen as a sacred gift from the gods.

3. The world is an organism, in which animal, vegetable and mineral components are equally alive and (a) invested with divine spirit, found in (b) the elements of Earth (sacred mountain, cave); Air (firmament, prayer flags, divine birds); Fire (comet, bonfire and altar flame); and Water (spring, river, lake and pool).

4. The entire cosmos and its divine creator(s) merge as a sacred dynamic creature, subject to the cycle of birth, maturity, decay, death, and re-birth. This macro-cycle is often envisaged in superhuman terms, supported by an anthropomorphic interpretation of local topographical features.

5. Myth is re-presented by community ritual, timed to coincide with significant solar, lunar, and stellar events. These are staged at earthly locations regarded as particularly favourable and numinous. Such rites typically involve visual symbols – idols – ranging in scale from figurine to planned *architectural* representations of the divinity, together with liturgical evocation by prayer, poetry, and sacred song.

There and then, superhuman and human beings meet and briefly combine, with the supernatural imprint regarded as the model for human behaviour.

6. Such ceremonies are often held during a night of 'vigil' before the start of hunting, gathering, or harvest seasons, because humanity follows but never precedes, what the gods are believed to have initiated, as witnessed annually during those festivals.

7. The union of opposites, (night-day, asleep-awake, internal-external, death-life), is acknowledged and celebrated. This is either a partial interchange, or a complete metamorphosis between species, resulting in humanoid-bird-horse-ox and vegetable amalgams, ceremonially shown and performed; or totemic identification with a divine animal.

8. Instead of the modern plethora of univalent 'signs', multivalent poetic symbols and ambiguous effigies often convey paradoxical truths.

9. Synthesis takes priority over analysis.

10. Abstract concepts and theorising are largely absent.

11. Subjective attachment is the norm, rather than a subject-object distancing.

12. Intellectual, physical and emotional realms are integrated.

13. Ancestors are regarded as participating influences among the living.

14. Collective identity takes precedence over human individualism.

15. Firmament, earth surface and underworld offer a vertical axis, up and down which the deities ride. Offshore islands provide Other World sanctuaries.

16. An appreciation of micro-macro affinity contributes to a widespread sense of belonging. The house is the working, divinely built universe in miniature.

17. The multi-layered intertwining of different 'divine realities' is often found. Myths can take compound forms after assimilating diverse (or mutually hostile) mythic strands, such as male-female rivalry, and in Pagan-Christian hybrids.

18. The supernatural pantheon is a template for human morality and social organisation, including the duty to maintain a balance between expenditure and resources.

19. Sacrifice (in gratitude or hope), as in tithes, or seed-corn planting underground, plus animal and human sacrifice, are aspects of a 'give and take' exchange.

20. As Gimbutas demonstrated in her *Journal of Indo-European*

Studies, mythic attitudes lie slumbering in the dictionaries of every nation. Recognising the family resemblances among cognate words in the Indo-European group of languages helps to arouse the 'sleeping beauties' of superficially 'lost' beliefs residing in all these tongues.

21. Vestiges of mythic beliefs and customs survive in oral and recorded folklore, and are potentially a valuable aid to archaeo-mythic research. Demonstrating the wisdom of the 'uneducated', the folklore residue is a pre-Socratic legacy, in which images 'speak for themselves' and contribute to the story-telling process. Similarly, prior to Plato's intervention, the Greek word for 'idea' was linked to the verb 'to see' and to the noun *eidolon*, 'visible image' or phantom. Whether as artefact, architecture, topography or zodiac figures, intellect and spiritual value lay within physical reality. Mind and matter remained united until the 'educated' were taught to cut 'idea' from 'thing', and convert ideas into Plato's abstract notion of *non*-material ideal 'forms'.[8] (Consequently, those academic departments of prehistoric archaeology who follow his lead do not expect to discover traces of the archaic mind in the matter that they excavate.)

22. By contrast, the combination of verbal and visual intelligence, including a folk player's costume, seen swaying to the pulse of folk song and dance, legitimately enlivens and marries a range of faculties. On performing or witnessing such an event the locked door to myth half-opens of its own accord.

23. From a mythic perspective, the most comprehensive and efficacious kind of 'functionalism' stems from Deity. Therefore the apparently mundane tools and vessels of 'ordinary' life are regarded as microcosms of a divinely made universe. For example, 'pot' = cosmic womb.

24. Whereas most modern scholarship is compartmentalised into self-isolating departments, the mythic world was (and is) characterised by interplay between the different aspects of knowledge. As the Institute has convincingly shown, in myth, the modern boundaries between art, religion, geography, farming, language, inherited place names, astronomy, and technology are absent.

25. Myth readily digests *historical* changes within its resilient cyclical rotation, thereby gaining additional energy and continuing relevance. Myth is typically a-historic.

26. The cities of the ancient world were founded on mythic co-ordinates, involving sunrise and set, apogee and nadir, all displayed in their street layouts. Whether or not they acknowledge this legacy, modern city-based archaeologists have inherited mythically ingrained urban prospects.

Then And Now, Now And Then

In her 1995 essay *Looking Around*, the art historian Lucy Lippard hopes that today's artists can help 'to reinstate the mythical and cultural dimensions to public experience.' She continues: 'We need artists to guide us through the sensuous, kinaesthetic responses to topography, to lead us into the archaeology and resurrection of land-based social history'.[9]

In myth, the repeatable dramas of the deities, played in the space-time of their dynamic metabolism, and re-presented on the holy stages provided by caves and ancient monuments, are at once archaic, repetitious *and*, like geological strata, ever-present. The unbroken mythic traditions of Hindu India[10] and the reviving Pagan movement in Britain affirm this truth, even while we all 'progress' in a straight line towards global disaster.

Myth continues to offer archaeology its valid original context; namely an inexhaustible drama, awaiting re-incorporation into the West's misnamed 'Past'. Thanks to Joan Marler's Institute, the reunion has already begun.

This challenge involves more than intellectual exercise, and more than an interdisciplinary 'objective' discourse between scholars, drawn from different fields. To rediscover the semi-hidden world of pre-Socratic *muthos* also requires a willingness to participate directly in the arts, plus a capacity to marvel and show gratitude for our planetary good fortune.

We need to switch off the light to see the stars, and merge constricted 'selfhood' into universality shared (it transpires), with those of an entirely scientific outlook; for they *also* believe (along with Einstein) that subject–object scrutiny is inevitably a two-way engagement. Moreover science now states that stardust is in our bones, while the water in our bodies is provided by rain and oceans.

The archaeo-mythic evidence suggests that the human species has, over time, found a variety of ways to express profound empathy with our native planet and with the grandeur of the universe. It is for us, even when apparently disenchanted, to find valid ways to appreciate and benefit from the compound wisdom of previous myth-based cultures.

MARIJA GIMBUTAS' GIFT

The prehistorian, linguist and folklorist, Marija Gimbutas (1921–1994) operated within the academic milieu, yet her interdisciplinary approach, and her emphasis on a goddess-centred view of Neolithic culture, came to be regarded as subversive by many of her fellow academics.

Marija Gimbutas used her command of multiple perspectives to dissolve the fragmentation into artificial categories by much modern research, and thereby opened the door to an alternative and long-lost world. From diverse frames of reference, she energetically developed a fruitful, interdisciplinary synthesis.

This enabled archaeology's 'things' to recover their rightful place within the mythic narratives that had originally inspired their production. Thus she devised and employed the term Archaeo-Mythology, to describe how excavated objects contained the clues to their former use in ritual and mythic narrative. She showed that silent pots and figurines were still transmitting an eloquent poetry; through a visual language that employed pattern, symbol and metaphor, to promote a calm integration of life's many aspects, including mortality, and to explore the partnership between humanity and other animals.

Scrupulously careful in her recording methods, she nevertheless operated beyond the airless confines of positivistic instrumentalism. In

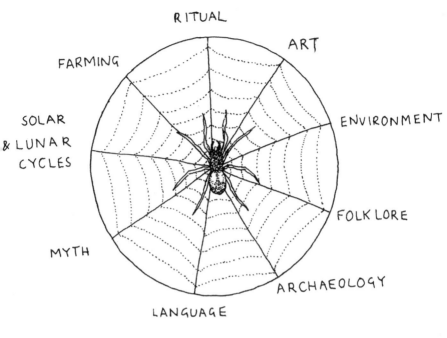

RITUAL
ART
FARMING
ENVIRONMENT
SOLAR
& LUNAR
CYCLES
FOLK LORE
MYTH
ARCHAEOLOGY
LANGUAGE

her challenge to prevailing orthodoxy, she was a true pioneer, who was obliged to leave others to fill in the details on the map that she was outlining.

With regard to language, she asked; if the pregnant Neolithic figurine squats under the tree of Indo-European speech, can comparable meaning-clusters be found in the proto-words of those vocabularies? In addressing this and related questions, she founded a journal, devoted to such speculations, and invited a wide range of experts to contribute articles.

Again, encouraged by her early experience as a collector of Lithuanian folklore, she asked: If Neolithic societies lived in close proximity to Nature's metabolism, did Nature's sacred heartbeat permeate their songs, dramas, dances and sculpture? And if so, may we not hear it today, despite superficial distractions? Are we really so far from the stations of night and day, that awe in the face of Nature now has no meaning for us?

When Marija Gimbutas died, she left us a legacy of hope-filled, but largely unanswered questions, and hinted that our enthusiasm for explanations should not stultify our ability to be and simply enjoy, in the presence of the ineffable.

In drawing attention to the linear patterns so often found painted onto Neolithic pots and sculptures, she gave us another starting point for prolonged contemplation. These red lines run like rivers and arteries to animate the static figurines as they trickle across the topography, from neck to thigh, bringing an inherent dynamic to these reposeful images. Whether read as plough-furrows, jewellery or wounds, (or all three together), these patterns open a discourse between surface and hidden interior, and between conscious observation and ancestral memory, outer dancing ground and core quiescence.

When allowed to perform ambivalently, the decorative web serves to both clothe and to undercut the single moment of regard, thereby extending it beyond the self, towards a chain of being and becoming, infinite in scope. Considering this prospect as one of Marija Gimbutas' gifts, how can I thank her adequately in words?

THE SILBURY MONUMENT AS A WORK OF ART, UNIFYING COMMUNITY AND COSMOS

In 1970, the English artist Richard Long ran a length of string from the top to the foot of Neolithic Silbury Hill. He then coated this string in white powder, before arranging it in a spiral on the darkened floor of an art gallery.[1] Hey Presto! The tallest monument in Neolithic Europe was reduced to a flat pattern, an abstraction to be appreciated on purely aesthetic grounds. This work of art was exhibited in the Tate and Hayward galleries in London in 1991, and then in New York and Los Angeles in 2009.

The artist commented with evident pride: 'I make my work as an individual. I actually hate the "Prehistoric Art" approach to my work'.[2] Prehistoric artists, all long dead, might have found his work inadequate on the following grounds:

1. His emphasis on his personal achievement, rather than making a contribution to communal necessity.

2. The exclusive exhibition of the work within interior, urban locations, far from the original rural setting.

3. The static, non-dynamic nature of his piece.

4. Its complete disconnection from the elemental and agrarian dramas.

5. Its failure to touch on the poetic interweaving of vegetable and human birth events.

6. His work's detachment from the First Fruits August Festival.

7. His lack of reference to the supernatural quality and purpose of the Silbury edifice, with its surrounding moat creating a gigantic water image of her body.

8. His failure to recognise Silbury Hill as the Great Goddess's bulging pregnant womb.

9. His artwork's failure to relate to the nearby 'birth spring' of Swallowhead.

10. His reluctance to consider any form of spectator participation in or around his artwork.

Perhaps Long's approach in this exactly matches the instrumental reductionism adopted by most modern archaeologists in their studies of that monument. For example, Long's carefully measured string matches their devotion to abstract numbers.

Yet in many respects the contemporary visual arts *do* overlap with, and provide possible points of re-entry, into the Silbury builders' mentality. For instance, some sculptors, not least Henry Moore, have treated topographical features in anthropomorphic terms, while many land artists, following Robert Smithson's lead, with his *Spiral Jetty*,[3] integrate their works directly into landscape. Moreover, recent performance art and happenings often invite audience participation.

Silbury's 'aesthetic merit' offers much more than 'good taste', since that monument shapes the world as a provisional fiction, displaying chaos resolved into a simple form. To the spectator, it offers the opportunity for an endless gathering and interpretation, of the fragments of individual experience.

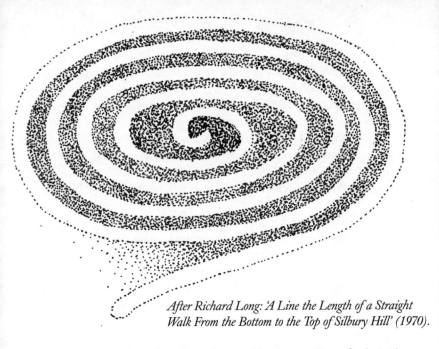

After Richard Long: 'A Line the Length of a Straight Walk From the Bottom to the Top of Silbury Hill' (1970).

Silbury displays the idea of a world (the totality of what is), embodied in a 3-dimensional arrangement of rock and water; across the surface of this liquid, every breeze plays, while at night the reflections of moon and stars replace that of the sun. These volatile events occur on a moat, deliberately shaped to register as the body of the Great Neolithic Goddess, shown heavily pregnant, with the Hill describing her full womb. In short, Silbury is a sublime creation.

According to Kant's definition of 1790, the sublime is the 'Absolutely Great', a boundlessness, located in the Sacred (the super-sensible substrate), underlying both Nature and thought.[4] Yet at Silbury this sublime range and achievement is combined with beauty, stemming from the shaping of its boundaries, into the goddess's reclining profile. Silbury integrates the sublime *with* the beautiful, by displaying the entire environment as a living, gigantic female body, whose metabolism includes the cosmos.

However, in our analytical age, her extraordinary feat of synthesis only serves to underline *her foreignness*. We now generally misread her as an uninterpretable misfit, hardly worth a length of string. But rather than confirming the superiority of our own culture, she continues to silently advocate the best interests of all humanity, by embodying a *dwelling at home*, in a universe full of sacred matter, freely shared. Thus, even when ignored, Silbury remains the greatest work of art imaginable.

It would be wrong to lose faith in the ability of contemporary artists to recover something of the Silbury spirit, as displayed by *Field* – an artwork conceived and described by Antony Gormley.

This 'sculpture' is composed of 335,000 humanoid clay figures, from 8–26 cm tall. They were made by members of the public. Here the clay as Earth carries the voice of the 'other', in order to reaffirm the spirit of the land as a continuum between mineral and human life. *Field* represents the invasion of cerebral purity by the chaos of Earth, and of western civilisation by the remote, the marginal, the dispossessed, and the unacknowledged. *Field* accommodates the 'spirit of the ancestors', the primal population, for whom mud takes on the attributes of sentience, and evokes those who are as yet unborn. The figures create a landscape of gazes, which look to us to find their place. Their eyes are the outward expression of insight. Here the heroic is put in the position of a child, with the mother being the Earth, with the child clinging on to her, in a form of expression that pre-dates Language.

Field expresses community, everyday humanity, where self-awareness is merged with the vast body of humankind. *Field* is a kind of soul garden, a collective creativity, extending into the boundless space of conscious, and a return of desire, joy in material contact, and the value of touch.[5]

Meanwhile, following his re-examination of Silbury Hill's core, the archaeologist Jim Leary concluded that 'people were deliberately bringing together the physical ingredients of their landscape such as chalk, river valley silt and gravel, sarsen stones and organic matter, as if to celebrate the land that they inhabited'.[6] And how can we forget that they were people of the Neolithic era, our first farmers?

To emphasise the annual range of farm work, Silbury incorporates into its base ground several concentrically ploughed furrows. These trenches can be interpreted as a collection of ritually dug 'first furrows', with each one initiating a new ploughing season, a habit continued into Classical times, when it was termed the *sulcus primigenius*. In 19th

Century English folk practice, this ritual furrow was dug on January 6th, known as Plough Monday, every year.[7]

Beneath Silbury, a massive, repeatedly widened trench, 100 metres in diameter, was also found.[8] This may well allude to the massive ditch that surrounds the nearby Avebury henge monument, which has been shown to evoke May Day hopes of future harvests.[9] In other words Silbury incorporates signs of previous seasons into the ground on which the harvest climax temple is built.

The cycle was further emphasised by the Silbury builders, when, at the time of that Hill's construction, they deliberately filled the chambers of the nearby West Kennet Long Barrow with layers of soil and vegetation, thereby integrating the ancestral bones which it contained into the agrarian cycle, while erecting an additional ceremonial forecourt to that barrow.[10]

In contemporary 'Earth art' terms, Silbury can be seen as a primal mountain, (it is the tallest prehistoric monument in Europe), which stands for the entire fertile Earth. But here that abstract function also features as the pregnant womb of the Neolithic Great Goddess, surrounded by the rest of her reclining body, defined by the surrounding moat, which depicts her in the squatting position adopted by pregnant women, as shown in hundreds of figurines from that era.[11]

Thus Silbury also functioned in terms of modern performance art, involving 'audience participation', both during and following its construction. Considering the dense pattern of human and animal footprints that Leary found beneath Silbury Hill's primary mound, he concluded that this particular patch of earth had been a ceremonial 'dancing ground'.[12]

In addition, the incorporation into the Hill's design of a wide ledge, encircling the Hill, 17 feet below its flat summit, allowed the people's representatives to be lodged on the hill, with a view of the entire moat image, and with close access to the field of 'primary fruits', the sacred grain crop, growing on the hill's summit, standing and waiting to be cut. Likewise a flight of steps found leading into and under the moat from the base of the mound, suggests another Hill-moat link, for communal use.

Hill and Moat together formed a sacred theatre, shaped into

the form of the presiding goddess, depicted on the verge of harvest, with the interplay of water and earth essential to all life forms, both animal and vegetable. Thus 60% of every human body consists of water. Here her 1,109-ft. (338 metre) effigy makes a 'Lady of the Lake', a work of art that *works*. As if to confirm the breadth of this mother's scope, and her deep attachment to the landscape in which she is set, every year until the mid-19th Century, water was carried by local people from the nearby Swallowhead springs, and drunk on Silbury Hill's summit.[13]

At Silbury, harvest, river birth and human birth were assimilated, and celebrated in the shape of corn dolly effigies, made from plaited straw, made in Mother Earth, Hag and Maiden shapes, in further recognition of life-cycle continuity. Ordinary farming was suffused with supernatural purpose and authenticity. Silbury is plainly an example of 'Religious Art' that bursts the boundaries of any narrow definition. Because the moat effigy is now all-but filled with 4.6 metres (15 feet) of accumulated silt, it nevertheless continues to be topped up with water, on which swans and ducks may often be seen swimming. This water is fed by a ring of minor springs around the moat's edge, where a layer of hard chalk rests on semi-permeable marl. Indeed, this geological circumstance helps to explain why Silbury was built in its low-lying, valley bottom location.[14]

The midpoint between June solstice and September equinox is the early August Quarter-day, signalling the start of harvest – called *Gwyl Awst* (August vigil) in Welsh, Lughnasa in Ireland, and Lammas or loaf mass by Christians.[15] As the Welsh name implies, it involved an all-night vigil, since birth events, both vegetable and human, were associated with the full moon. On the night of that full moon at Silbury, it rose to throw its light on the water deity's thigh, at a point where her new child's head would be seen to emerge. Then the moon would swing to the south side of the Hill, to illuminate the dry Cradle, the only dry section of her moat. These beams would simultaneously align with the Swallowhead springs, before, as dawn approached, the last moonlight would set over the water figure's clearly defined breast, so delivering the first milk to her new-born child.[16] In our terminology, Silbury is a kinetic artwork.

With the return of daylight, the Hill's inverted reflection, entirely visible in those waters, completes the hill image into a nearly-spherical entity, in which her underworld presence provides the support, upon which surface reality is seen to float.

Meanwhile, on each cloudless night, planets, stars and entire galaxies, swam in and decorated her moat. The entire canopy of the visible cosmos came to her, in a universal reunion, an upper field of golden seeds to complement her earthly harvest of ripe wheat. As for the solar system, since the moat's long axis is aligned due east-west, in 12 hours, it gives sunrise from her crotch and, precisely 12 hours later, sunset over her head, equal measures of light and dark. What could be more reasonable?

Silbury confirms architecture as 'the Mother of the Arts'. In her temple, there was room for thousands of people to stand around the edge of her moat, where they could see their own reflections intermingled with the endless recycling of everything, in a numinously enlarged version of reality, past present and future.

As Charles Jencks has recently remarked, 'prehistoric art can bring many values together in a cosmo-genesis, and in a way that many goddess based cultures instinctively understood'.[17] The great 20th Century artist Paul Klee spoke of 'that secret place where primeval power nurtures all evolution... the womb of nature, at the source of creation'. He asks: 'who would not dwell there?'[18]

In a calmer vein, the philosopher John Dewey declares that 'The function of art is to organise experience with more than usual meaning, vividness and coherence'[19], while Nietzsche considered that 'Art is the affirmation, blessing, and deification of existence. As the deliverer of plenitude, it encapsulates a desire for stability and Being; while accepting Death, Change and becoming'.[20]

Similarly, Tolstoy concluded that 'if a society truly lives, the religious feelings transmitted by the artist through sincerity, clarity and intensity enables religious perception to operate in actual life'[21]; while Jung noted that 'the entire creative process has a feminine quality... we might say from the realm of the Mothers'.[22]

So the multitude gathered around the Silbury moat-deity at *Gwyl Awst* hoped to enjoy a pagan incorporation into the event. At a given signal, Splosh! they could all jump into the moat to demonstrate immanence enacted, so completing what Coleridge named as a work of the Primary Imagination, 'A repetition in the finite mind of the eternal act of creation, in the infinite I AM'.[23]

SILBURY AT PLAY:
THE GODDESS IN FOLK DRAMA

The German philosopher and sage, Martin Heidegger, never set eyes on the Silbury monument in Wiltshire. Yet there is much in his later work, between 1935–1959, which sheds light on temples in relation to the Earth, and as articulated by Silbury.[1] He considers the basic purpose of all temples is to link humanity to the gods, as a result of joining Earth to Sky, in a four-fold grouping including humanity and the divine.

The temple's foundations in the underworld are united to the earth's surface, and from there to the sky, including sun, moon and stars. By means of this four-fold grouping people are able to truly experience an essential 'oneness' with creation in general[2] – and can thereby experience a sense of 'dwelling' on this planet, in the fullness of its offering.

So, in standing there, 'the Temple first gives to things their look, and to people their outlook on themselves'. The temple's radiance amounts to a 'letting lie forth', and its boundary is a 'setting free into the unconcealed' of 'a repose charged with the fullness of motion'.[3]

At Silbury, where the temple's boundary defines the water and earth image of the pregnant Goddess, we find 'the fixing place of truth in that figure'; namely the union of Culture's immanent and repeatable harvest with the natural endowment from which it grows and on which it depends, envisaged as the repeated gift of the Mother Goddess, the fertile Earth presented as the pregnant, upstanding womb of the Great Goddess of the 3,000 year-long Neolithic era.

Heidegger and Lalitha

Heidegger welcomed the rich legacy of alternative views of reality inherited from earlier civilisations. He understood their latent relevance to contemporary life. To him, 'Tradition, at its most profound, is eternal incipience. Its transmissions have an inexhaustible give... Out of the things themselves can come a hidden treasure of fullness, the wealth of the simple, the well-spring of the ineffable, an aboriginal gathering of essence, and the dynamic stillness, that is the source of history, so old, yet ever so new.'[4]

Thus at Silbury, the Earth in its *entirety* is re-presented as a fecund gift of the pregnant mother deity, with her womb, flattened on top, ready to accept the next token field of growing corn, as the agrarian symbol of the country-wide annual harvest child, coupled with the gigantic mother's simultaneous delivery of a ground-based human infant.

Just such a revelation can be found at Silbury, where the lake that surrounds the 130 feet high hill was deliberately designed in the shape of a pregnant squatting mother goddess, chief deity of the Neolithic era throughout Europe. Completely surrounding Silbury Hill, she is both Water and The Earth into which she is grounded, while her pregnant womb is described by the hill, arising from her watery body. (Ten feet of accumulated ditch silt, together with a fall in the water table, now spoil the show, in some seasons.) That the long axis of the water figure defines the due east-west line of equinoctial sunrise and sunset, lends an additional stability to the work in progress.

At Silbury, rock and earth is literally raised up from normal mundane matter, to share in the goddess's divine, nurturing metabolism. So the Hill stands, ready to deliver the next harvest child, from the sacred ground of her being.

Silbury, in Heidegger's terms, is a 'Poetic Image', which offers 'imaginings as visible inclusion of the alien (such as distant stars) in the sight of the familiar (such as a maternal human form), gathering the brightness of heavenly appearances into one with the darkness. By such sights the deity surprises us, and proclaims her unfaltering nearness'.[5] As Heidegger says of such monuments, Silbury 'displays repose in the fullness of motion'[6].

The 130-feet-tall Silbury monument, seen from the East.

At present, British researchers continue to regard the precious moat figure only as an unfortunate by-product of the hill-building process, or, at best, a defence against evil spirits;[7] ignoring the world-wide positive value placed on water, as outlined in Mircia Eliade's majestic survey.[8]

'Water symbolises the whole of potentiality; it is *fons et origo*, the source of all possible existence, and the primal substance from which all forms come. In cosmogony, in myth, ritual and iconography, water *precedes* all forms and *upholds* all creation. Emerging from water (as does Silbury Hill) is a repetition of the act of creation in which form was first expressed'. It is a 'new birth' because emersion fertilizes, and increases the potential of life. Eliade goes on to say that 'since prehistoric times, water, moon, and woman, were seen as forming the orbit of fertility, both for man and the universe', with 'water viewed as the universal mother'.[9]

But for Silbury to fully display architecture as 'Mother of the Arts', she needs to be *occupied* by humanity, raised to a fully animated condition by human traffic; and Heidegger sees such human gatherings in terms of ritual dance.

He writes of the fourfold unity resulting from; 'the appropriating mirror play of the betrothed, each linked to the other in simple oneness – the round dance, the ring that joins as it plays, into the simple one-fold of Worlding'.[10] This involves replacing a thinking that merely represents and *explains*, with a thinking that responds and recalls. Silbury provides an architectural stage that accommodates such performance in the interests of communal cohesion, and caring.[11]

Around the edge of the Silbury moat-image, there is room for a chain of at least 1,500 dancers, holding hands, and space for another 500, standing where the hill and moat meet. Here humanity could demonstrate Heidegger's 'four-fold unity' as a matter of fact, and in doing so, they could find themselves measured against infinity, in a truly religious experience. He describes this as 'a letting come of what has been dealt out' and 'a gathering-ness into an original being-at-one'. The dance leads into the radiant world of the *Ereignis*, the Festival.[12]

So Silbury, built to display repose combined with motion, reached to the edges of the firmament, to grant humanity a full sense of 'dwelling', involving guardianship, and an understanding of the holiness of everything. In this approach, the 'because' of rational explanation is swallowed up in play. 'Yet the play is without a "Why?" It remains just play. But this "just" is everything, the one.'[13] Silbury, by standing there, can still 'give to things their look, and to humanity its outlook, for as long as the deity has not fled' from the monumental moat.[13]

The notion of deeply serious play has almost vanished in modern Europe, although traces linger on in the carol or round dances of our fading folk tradition. By contrast, India has maintained the 'play' tradition within worship, inherited from the Stone Age. Thus the Indian word: *lila*, 'play', is embodied in the goddess *Lalitha*.[14]

In Hindu tradition, Lalitha's name means 'She who plays, spontaneous, easy',[15] while *lila* refers to the creation and sustaining of the universe.

In Lalitha's games, she combines *vaya* (air), *jala* (water), *agni* (fire), *bhumi* (earth) and *akasa* (space). As mistress of this annually re-enacted elemental sport she is the primordial shakti, who existed before Brahma and the gods, who is believed to bring peace, knowledge (vidya) and wealth. As *Jagadamba*, Mother of the World, she is the ultimate female power, and in her triple manifestation, combines Motherhood, with Daughter and Hag roles, (just as Silbury is flanked by, and stands within sight of, both the marriage ring, seen in the Avebury Henge, and Britain's greatest Neolithic long barrow, sited nearby at West Kennet).[16]

As the all-sheltering Mother, *Jagadamba*'s breasts are likened to mountain peaks, which give birth to rivers. Similarly, the Silbury mother's placement probably relates to the nearby Swallowhead springs, source of the river Kennet, a major Thames tributary.

So, for both Lalitha and Silbury, creation means 'to fetch from the source, and to bring what has been so received'.[17]

The tract known as *The Lalita Sahasranama*[18] contains the goddess's thousand names, arranged in hymns, where the metre is strictly observed. It starts by calling her Shri Mata, (Great Mother), then Shri Maharagni, (Great Queen), and describes her as 'the one who is as bright as the rays of a thousand rising suns', and the 'one who is holding the rope', and 'one who was born from the altar of the fire of consciousness'. Her home is in Nagara, a city built on top of Maha Meru, the mystical mother mountain.[19]

Attitudes embodied in Lalitha and *lila*, can be found in archaeological evidence from the Neolithic Indian cities of Harrapa and Mohenjo-daro, (including circular ritual enclosures, and images of dancing girls), made before 2000 B.C.[20] One can legitimately say that, across the intercontinental span of the Indo-European language, *lila*, 'play', was part of the common currency of worship, stretching from the Bay of Bengal to the Silbury monument. And at both ends of this range, the underworld, the surface plain, and the celestial vault, were integrated and jointly honoured by means of playful worship.

But because this 'play' is a matter of life and death, it includes a sombre aspect. Thus Silbury stands within sight of the West Kennet Long Barrow, where the deceased were assembled within a stone-built image of the deity, in her mistress of the dead mode. And as a letter relating to the 1770 excavation of Silbury makes plain,[21] a thin, vertical, deep, natural water-worn shaft, found beneath Silbury's core, was said, in local folk belief, to connect with the long barrow, as directly as summer joins winter.

In addition, the carefully placed layers of soil and organic material – found covering the barrow's thousand-year-old bones during Piggott's 1956 excavation of West Kennet Long Barrow – is believed to have been inserted at the time of Silbury's construction, (as indicated by mid-third millennium pottery fragments, included in the overlying deposit.)[22] This points to a determination to sustain the life-death cycle at both ends. The long barrow was not a cul-de-sac. The annual dance included it, but did not stop there. And Silbury Hill's inverted reflection appears in the moat, to complete its partnership with the underworld.

In the context of Heidegger's thought, Silbury provides a complete sense of 'dwelling'. For him, this involved 'preserving the fourfold of earth, sky, divinities and mortals, in its essential oneness, fostered by human care'. By building the Silbury monument, the community achieved that sense of dwelling and 'brought it to the fullness of its nature, thereby giving access to the great whole of the unbounded, seen emerging from, and united with, their native ground'.

III.
The Cyclical Goddess

THE CAILLEACH IN GAELIC
SCOTLAND AND BEYOND

The Cailleach is the all-important Hag of Winter, sacred crone or dark goddess of Scotland, Ireland and the Isle of Man. Her function is to return the isolated human species to the realm of Nature, and to re-establish the fading sense of connection between humanity and all other animal species.

With her spirit the Cailleach reanimates the elements of Earth, Air, Fire and Water and infuses the plant realm, in both its wild and cultivated aspects. She enables the free exchange and intermingling of properties between them, so creating a sense of unity and environmental synthesis.

She offers each person a sense of 'being at home' in the cosmos, with the individual operating as a microcosm of the universe.

She is the agent of continuity, establishing a poetic fusion between human life thresholds, the seasons and agricultural phases, in a polyphonic achievement, further secured at seasonal festivals.

She thereby coverts linear, historical time, and concepts of linear 'progress', into cyclical, repeatable, myth time.

The Cailleach is international in her scope. She is related to the skull-wearing Hindu goddess, Kali; like the Cailleach, she also has a sky-blue face.

The Cailleach brings renewal from winter's abyss, and so she often operates below ground level. She personifies the extremes of pleasure and pain, resolved into harmony.

Her winter season stretches from Samhain to Beltaine, overlapping with that of Brigit's feast of Imbolc by three months. Likewise, her name, Cailleach, 'hag', closely resembles *Caileag*, meaning 'Maiden'.

The Cailleach offers a mind-body reunion, and a sacred world view, free from guilt. She invites both men and women to a celebration that preceded, and may outlive, the 2,000 year-long Christian interlude.

The hooded crow, Corvus cornix, *a harbinger of death.*

The Cailleach can operate inside households. For example, my wife often manifests as the Cailleach; old, fierce, frightening, loving and loveable.

In scale, as a working symbol, the Cailleach can range from an embodiment of the four elements – Earth, Air, Fire and Water – to the topographical totality of a country and a continent – including its mountains, valleys, plains and coastline, plus its lakes and rivers and streams, snowfall, rain and wind, meadows, forests, birds, deer and goats, cattle, horses and crops – all flourishing under her blue-faced sky, and warm, light-giving sun; to the internal physical fabric of the individual, including its skeleton, liquids – typically 60% water – flesh and bodily warmth of the living organism.

The Cailleach's spirit tends to be embedded in matter. Therein she resides, to animate and embody a blessed world.

Gaelic Scotland, especially the Highlands and the Western Isles such as the Hebrides and Skye, are heavily influenced by Irish settlement and speak a language closely related to Erse. Their language contains the supernatural figure of the Cailleach, the super-natural hag who is believed to preside over the winter half of the year, from November to April. She was termed Beira, Queen of Winter.

In Ireland, her full name, Cailleach Bheure, may link her specifically to the Bheur peninsular in Co. Kerry, yet in professor G.O. Crualoich's opinion, that place-name derives from the word *beurach* meaning 'shrill or sharp'.

Prior to the Cailleach's arrival in Scotland, a mother goddess had been worshipped in Neolithic times, from c. 4,000 B.C. onwards, as stone figurines, found in the Northern Isles, demonstrate. A matriarchal culture was also developed by the Pictish peoples of northeast Scotland, which was likewise based on goddess-worship, in tune with the solar and lunar cycles, and with the human and farming threshold events. This poetic fusion of recurring changes, supernaturally determined, operated beyond the framework of tribal or national boundaries, and of historical periods, and prevailed into recent times. D.A. Mackenzie remarks: 'A remarkable feature of Scottish mythology is the predominance of goddesses. They are greater and stronger than the gods. They often take the form of *The Cailleachan Mor*, "the great Cailleachs."'[1]

Yet on both sides of the Irish Sea, the Cailleach operates largely within a triune structure of inter-connection, however isolated she may appear in the depths of winter, old age, and physical decay. She is perpetually teetering on the verge of transformation into her maidenly or maternal modes. Consequently, in Scotland she is often regarded as a complementary aspect of the goddess Brigit, alias Bride, with whom she overlaps in time, between February 1st, St Brigid's Day, and May 1st, Beltaine, when the Cailleach 'turns into' or at least gives way to, The Maiden. Needless to say, during her early spring arrival, Brigit takes on some of the Cailleach's underworld characteristics. For example, she is said to emerge from the *knoll* (Irish *sidhe*), on February 1st, in serpentine form. Across the Isle of Man, on that day, if the Cailleach was seen as a gigantic bird, carrying sticks in her mouth for her future firewood needs, people guessed that the remainder of winter would be long and hard.

Scotland, we should recall, lies nearer the North Pole than both the Isle of Man and Ireland, and has a much more mountainous terrain. Consequently, there the Cailleach was very reluctant to give up a single day of her six-month rule. She was still going strong on March 25th, *Latha na Cailliche*, 'The Old Woman's Day', when the struggle between the forces of winter and spring were at their most intense. It took place around an alder tree, from which whistles were made, that could summon up the spirits of dead kings.

By that date, the Cailleach was said to have captured Brigit, but Aengus, Brigit's brother or suitor, managed to extricate her from the Cailleach's clutches.

However, the Cailleach's power remained formidable, including: *Feadag*, 'The Whistler', namely three days of plover-like wind blasts, followed by three days of piercing east winds, called the *Gobag*, 'the sharp-billed one'. After that came the *Sguabag*, 'The Sweeper', three days of *soughing* (murmuring) blasts, offering a hint of springtime, that was soon confounded by the *Gearran*, nine further days of tempestuous weather, which bucked and reared like a colt. Finally, in The Old Hag's week, of roaring ocean waves, the Cailleach milked her goats. (Seen in the mythic Glastic figure, the Cailleach could appear as half woman, half goat.) But eventually, she gave up her attempt to suppress the newly rising grass with her staff, so she threw it under a holly tree, where grass never grows, and with a scream she then admitted defeat, by which time it was mid-April, and her half of the year was all but over.

In some districts it was believed that the Cailleach turned to stone on May Eve, only to regain human form at Samhain. In southern Scotland, her stony summertime persona is recognised in the many 'Carlin Stones', established since prehistoric times, as at Commoncraigs, Ayrshire; and at the Carlin's Tooth rock outcrop; and a dozen other sites, including the 30-feet-high pillar, named Carlin Maggie, overlooking Loch Leven, Perthshire; and The Hag's Stone, at Dunlop, Ayrshire. Other examples were formerly part of a prehistoric cairn or were derived from a prehistoric stone circle. At these places, the Cailleach returned both to her ancestors, and to her original home within the nation's geological fabric. At Taynuilt, Argyllshire, there is a 63-feet (16 metre) tall pillar, rededicated to Lord Nelson in 1805. It had been carried there by the Cailleach from Upper Loch Etive, but the withies, tying it to her back, broke, so she dropped her load at Taynuilt, where the great stone took an historical turn. Yet given its origin, one might say that here the Cailleach infiltrates and undermines simple history by her timeless persona, despite the claims made on the pillar's modern label.

Other accounts say that instead of petrification, on May Day the Cailleach recovered her lost youthfulness by bathing in a certain loch, or that she took part in the Beltaine procession as an old woman, (with her role played by a man, cross-dressed), accompanied by a hobby horse and another man, dressed in black. On the Isle of Man, there were two rival Beltaine processions, one headed by an 'old woman', the other by a 'maiden'. They then 'fought' one another in a ritual display of violence.

Often gigantic in size, the Scottish Cailleachan Mor personified the forces of Nature. Tempest, thunder storm, the raging ocean, and flooded rivers, that annually sought drowned human victims. All aspects of the harsh environment were seen as manifestations of these furious underlying spirits, who had to be propitiated with offerings, or charmed with magic ceremonies, before their violent energies would subside.

Thus the Scottish Cailleach, in her various environmental guises, played across the rugged, storm-wracked land, that she was credited with building. Some folk accounts derive her from Scandinavia, rather than Ireland, (perhaps displaying relatively recent Viking influence). From there, she is said to have carried an immense creel of material, that leaked through its broken base, and so formed Little Wyvis, and other mountains, including Argyll's Ben Cruachan, (alias The Hollow Mountain), and An Cruachan, on Skye, plus Creag na Caillich, near Killin, Perthshire, on which a Neolithic axe factory has been discovered. In addition the pyramid-shaped Ailsa Craig island, is claimed to have been accidentally dropped by 'witches', on their way to Ireland. As for the extent of her mountain-building role, the Cailleach was often said to have created the entire country, with many Highland parishes having their own version of her. She is also credited with activating innumerable streams, named *Allt na Cailleach*, in her honour.

In one such *Allt na Cailleach* glen in Perthshire, there is a small stone hut, re-thatched annually, in which her stone image, along with images of her Nighean, 'daughter', and her 'old husband', the Bodach, reside. These effigies are brought into the open and displayed from May Day till Samhain, when they are replaced within their shelter for another six months. Legend says that the Cailleach and her family were welcomed into this glen, and that while their images remain there, it will be fertile and prosperous.

These idols condense and concentrate the entire range of their subjects' powers, and their survival, despite centuries of Protestant iconoclasm, is remarkable.

The Cailleach is also recognised in the winds and gales that so often blow over the Scottish landscape, as in the Moray Firth, lying towards the North-East of the country. Here, ironically, she is known as Gentle Annie, and she is particularly feared by fishermen, though her name may be derived from Scottish fairies, often referred to as 'peace-folk', or 'good-folk'. Here she is the hag of the south-west wind, particularly feared in a stormy period in early spring, lasting up to six weeks.

During this time it was impossible to put to sea, such was Gentle Annie's fury around the Heel of Ness, where the wind, blowing through a gap in the mountains, and combining with a strong-running tide, made a lethal combination, that was challenged by 'ceremonies of riddance', which included lighting bonfires on the hills.

Off the Western isle of Jura, in the notorious Corryvreckan Whirlpool, the 'Cauldron of the Plaid,' was where the Cailleach was said to wash her plaid shawl for three days, in early winter. Then, when it had turned pure white, she pulled it ashore, and spread it across the land as the first fall of snow. So, her domestic chores had a country-wide effect. She combined sea with air-born snow and laid it on the land, in an elemental union.

Rather than the epitome of alienation, the Cailleach often joined things together. The formation of many of Scotland's mighty lochs were also attributed to her action, or, as in the case of Loch Awe, to her weariness in forgetting to control the well at its source, having just spent the day herding deer, another of her winter tasks.

Her close association with animals also included the hare, numbers of which often took refuge in the last patch of corn to be cut, while another of her names was *Gobhar Bhacach*, 'The Lame Goat'. In Wales, she is synonymous with the Harvest Mare, and, as in Scotland, she could often take the shape of an owl, *Cailleach Oiche*, or that of a cormorant, *Cailleach Dubh*, or even a woodlouse, *Cailleach Crainn*. Plainly her powers stretched well beyond the human species. On the Orkney Isles, she was recognised in the last sheaf of the harvest in bitch dog form. This straw image was tied to the farm gate, and the last man to enter the yard was pelted with clods of earth, and barked at, for he was to be her consort during the following year.

The appearance of the Cailleach at the end of Harvest ceremonies, that in Scotland usually took place in late September, seems strange if she was supposed to have disappeared during the summer half of

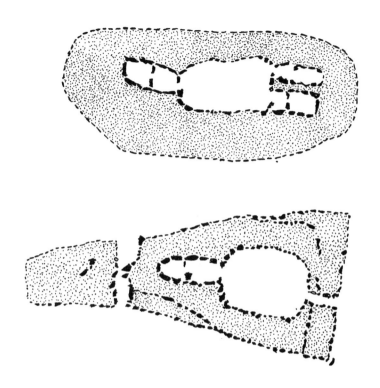

Neolithic graves at Creevykeel and Deerpark, Co. Sligo.
The Mother providing a home for the dead.

the year. Is this another instance of her 'overlapping' with Brigit's work? Or should one also argue that the cutting of corn is a form of vegetable 'death', that provides a 'straw bed' onto which she might comfortably recline? In fact, the Cailleach was widely-recognised to have been in the growing crop, throughout its process of ripening. She is 'the spirit of the corn', its fecund essence, and humanity's 'staff of life' from its seedtime inception onwards. Consequently, in Scotland, the 'corn spirit' was recognised and named both as Cailleach and as 'The Corn Maiden of harvest'.

Consequently, in Fife, two 'last sheaves' were cut – with one named the 'Maiden', the other the 'Old Woman'. There the oldest person in the field did the cutting and the youngest the binding of these final sheaves. Thus the Cailleach's effigies spanned the generations and the years, so linking the current harvest to the next.

She also crossed the female-male divide, in that the wheat seeds contained in the end-of-harvest effigies are termed 'Kern', with the same word denoting male 'Peasant', since he was an important part of her workforce. In South Uist, the Cailleach image was given slippers to wear.

The effigies displayed at the harvest supper were carefully dressed, and treated with reverence, for the corn spirit resided in them, and contained the concentrated spirit of the entire harvest. With time mainstream culture has forgotten the importance of these icons, and in recent times in England they have been relegated to 'corn dollies', a mere child's plaything. Yet this 'dolly' term is undoubtedly reduced from 'idol', the corn spirit as Deity, inherited in an unbroken line from our first Neolithic farmers, c. 4000 B.C.

The potency of the art object also applied to the crafted cauldron. In pre-history technical skill and 'magic' (craft) were often combined, as opposed to the modern tendency to separate art from craft. For example, in Wales, the Nature Goddess named Ceridwen, whose name has been variously translated as either 'the hunchbacked witch', or 'the beloved white one', brewed all the herbs of the year in her cauldron, believing that they would produce the elixir of wisdom. When three drops of this liquor accidentally splashed onto the youth Gwion Bach, he was eventually transformed into Taliesin, 'Shining Brow', the youthful sun god, fished from the sea, on May Day. Here as elsewhere the cauldron functioned as a useful image of the underworld, a reservoir containing an epitome of the earth's surface values, from which another year might be born.

In modern times after the Scottish Harvest Supper, the corn dolly/ effigy was usually hung in the barn, and eventually fed to the horses at the start of the next season's ploughing. Thus the continuance of the corn spirit's favours was assured. In the Hebrides, it was the custom for young men to pick an oat cake from the Cailleach's bonnet. The man who chose the burnt cake was treated as a scapegoat, and driven away from his friends, as a victim of her lethal fire.

The impact of farm mechanisation, and of Scottish Calvinism, combined to reduce the status of the Cailleach, but in 18th Century Northumberland she continued to be regarded as a Goddess, akin to Roman Ceres, and as the Harvest Queen. Accordingly, the last sheaf wore a white robe, multi-coloured ribbons, and the effigy was crowned with flowers. 'She' also carried a sheaf under one arm. So dressed, she was carried out of the village on the morning of the last reaping day, accompanied by music.

In the field, she was raised high on a pole, while the reapers held hands to make a ring around the last corn to be cut, and so linked together, they walked around the final patch seven times. Then, in the evening, they threw their sickles at these remaining stalks that carried the last wheat-ears. When cut, they plaited these into the effigy which they carried to the harvest feast. There she was raised on high, over the whole company, who curtseyed, danced and sang around her.

In some parts of Scotland, the last crofter in the district to complete his harvest had the onerous duty of keeping the Cailleach or Carlin image safe through the following winter. So she maintained her central role in the religious and physical well-being of the entire community.

If anyone tried to throw the Cailleach image on the house fire, he or she was always prevented from doing so, and then, as a penalty, that person was obliged to leap over that fire three times.

In English folk dramas, the Cailleach and her younger manifestations are fully integrated into the farming year. For example, the Lincolnshire Plough Jags assemble to perform on Plough Monday, January 6th, at the start of the ploughing season.

The plough boys serve as oxen and pull an exceptionally large plough around the village. Then The Maiden appears and she must choose a mate. She rejects a number of potential partners, in favour of The Fool, who is nobody and everybody. But this clown, dressed in patches, is able to play every male part.

At the end of the play, the Maiden reappears alongside and as Old Dame Jane, who carries the Maiden's summer-born child, in the form of a corn sheaf. (The Fool denies being its father.) Old Dame Jane, otherwise known as Dirty Bet, absorbs all the characters and the completed farming cycle into herself. She carries a broom, with which she vigorously sweeps the remains of the old year away. Neither

Dirty Bet nor The Cailleach should be seen as isolated figures. On the contrary, they both serve as the embodiments of a vital synthesis, the fruitful resolution of extremes. Traces of their necessary presence have survived in many cultures.

A comparable flexibility occurs in Wales, with regard to the horse deity Rhiannon, whose name is believed to derive from early Celtic *Rigantona*, meaning Great Queen. In the Welsh medieval *Mabinogion*, Rhiannon gives birth to a foal and to a baby boy, delivered on May Eve. In a prehistoric bronze effigy in the British Museum she is depicted seated, surrounded by foals, a wheat sheaf and a plough.

At the end of each harvest in Wales, she reappears as the last sheaf to be cut, named *Caseg Fedi*, 'the End of Harvest Mare'. This three or four-legged plaited corn effigy is cut by all the reapers, aiming their sickles at her roots. When freed from the ground, the reaper cries 'Early in the morning I got on her track. Late in the evening I followed her. I have her. I have her!' The rest ask in a chorus: 'What do you have?' To which he replies: *'Gwach! Gwach! Gwach!'* (A hag, A hag, A hag.) So the noble mare assumes her winter-hag or Cailleach mode. 'She' is then carried to the farmhouse, concealed beneath a man's clothes, while the women of the house try to drench the effigy with buckets of water, and strip him of his clothes to find the mare. Then harvest supper begins, including the Rhibo game, in which a man and a woman are tossed into the air together, by two rows of six men, linking arms to form a 'tossing basket'.[1]

So the years and ages go round, under the Cailleach's veiled guidance. While in the far eastern background stands the terrifying, but life affirming figure of the Hindu goddess Kali, wreathed in skulls. Like Kali, the Cailleach was said to have a blue-black skin. The Kali may in turn derive, as Hermann Gundert thinks, from the Proto-Indo-European goddess named Kolyo, 'The Coverer', who was half maiden and half monster.[2] The Cailleach has an intercontinental reach.

The Cailleach may now stand 'beyond the Pale', set up by a combination of urbanism, (involving the disconnection of the population from the agrarian round), with that gap further emphasised by an academic methodology which normally favours objective detachment as the accepted route to authentic study, rather than as a face-to-face encounter involving emotive experience. So how can the Cailleach survive in these circumstances, given her enigmatic, paradoxical nature, and soil-based habitat? How? Because we still live under, and depend upon, and share the seasonal round with her, and we may even still enjoy the Cailleach's nightmares, along with the wisdom that she continues to transmit, as a repeatedly proven way to reconnect with, and to help sustain, our planet.

Some Cailleach place-names in Scotland
There is a rich abundance of Cailleach place-names in the Highlands and Islands of Scotland. Numerous mountain streams are attributed to her, as are many of the valleys through which they flow, along with the deep circular corrie lakes, situated among the mountain peaks, from which the streams flow forth. These represent the Cailleach's winter hideaways, to which she retreats, after the herders have left their summertime upland pastures. The Cailleach's name is also attached to many of the upland hilltops, variously termed *Beinn, Cnoc, Meall*, or *Tom*. She can also feature in artificial mounds, named *Cairn*, or *Casteall*; and as a cliff, *Creag*; or a big valley, *Strath*; or a small loch, *lochan*; as well as the 'nose', or *sron*, of several coastal features; as well as some Forests. She likes dark places.

On the mountains her tasks included herding the winter flocks of deer, and caring for the goats, both wild and domesticated. She is synonymous with the lame grey goat.

Near Cape Wrath in Sutherland, she is linked to a phallus-shaped coastal stack named after her 'old man' or Bodach. (The term derives from *bod*, 'penis'.) He shares her shrine house, near Glen Lyon. (He can also take the forms of 'snowman, scarecrow, codfish and Santa Claus'). Meanwhile, the Cailleach is credited with creating the notable line of scree stone that runs west from the Juras. That line is termed her *scriobh*, or 'writing'.

Across southern Scotland, the Cailleach appears as the 'Carlin', a term given to 12 of her widely distributed standing stones and other notable boulders that have featured in seasonal folk ceremonies. In Adamnan's *Life of St Columba*, he reports that St Columba expelled an innumerable host of malignant spirits from our island, when he saw them assailing himself; but he was not entirely successful. To this day, Scotland remains in the grip of the Cailleach, as a living power.

As her place-names make clear, she underlies, and frequently emerges from, the rugged landscape, not least when she is facing the breast-shaped mountainous 'Pap' of her younger self, (her Caileag), as she does across Glen Coe. In Scotland, the Cailleach is referred to as 'Beira, Queen of the Winter'. The term *Beira* may derive from the Gaelic word *beir* 'to take'. She takes the falling leaves and grasses and buries them during the long harsh winter; in the people's annual experience, she is truly very much alive.

Cailleach in the Dictionaries

As a storehouse of ancestral attitudes and of interconnections now all-but forgotten, try reading the dictionary, where the word Cailleach is said to derive from the Latin *pallium*, 'cloak'. Her cloaked image is the embodiment of winter. In Old Irish, she is defined as 'the veiled one'. Similarly, she appears in one corner of several Roman 'seasons' mosaics that have survived in Britain. Hooded and wrapped up against the cold, she holds a bare branch, yet glances towards a youthful Spring sitting in a neighbouring corner (Gaelic *caileag*, 'young woman, girl'). As *cailleoir* (calendar), the Cailleach keeps her eye on the whole year.

Moreover, Cailleach extends her withered arms to embrace members of the animal kingdom, including some birds, fish, and insects. She has all-life responsibilities. Thus she is Cailleach *breach*, 'the dog-fish', which doubles as Cailleach *traile*, 'a sea shore nymph'. She also features as 'the eel-like Peter nine eyes', which is found under strand stones, and is seen as Cailleach *ruad*, 'the loach'.

Among birds, as well as being known in the guise of Cailleach *oiche*, 'the owl'; the Cailleach is known in the form of Cailleach *dubh*, the sinister 'cormorant', plus the red-legged 'chough', of the crow family. In earth, she becomes Cailleach *crinn*, the humble 'woodlouse' and Cailleach *goile*, the 'tapeworm'. As Cailleach *ligeog*, she is either 'a dangling thatch weight', or 'a stone boat anchor', 'an over sprouted potato', or finally 'a fir stump found in bogs'.

Yet in the midst of harvest she can reappear as Cailleach *dearg*, a 'red poppy or corn rose'. Her negative aspect appears in *cailltech*, 'ruinous, destructive and in *caillemnach*, 'worn out', and in *caill*, 'damage, trouble', while functioning positively as *caillide*, the primary 'spirit of the woods'.

As for the future, she serves as *cailleoir*, 'a supernatural prognosticator, soothsayer', and as *caille cartai*, 'fortune teller' of human destiny, who is seen again in miniature as Caille *ligeog*, 'a doll', perched on the threshold between generations and seasons. No wonder that we feel inclined to celebrate the Cailleach. Whether as witch, hag, crone, or old woman, she is a formidable but healing creature of the collective imagination, continually proposing a re-united world. Her stern outline can be compared to that of a deciduous tree in winter, displaying many seasons through its array of twigs, branches, boughs and trunk, all reaching for their share of the light, yet all dependent on hidden underground roots. And at the tip of every twig is the red, fingernail-shaped, bud of renewal.

This chapter is based on a paper 'Exploring Old Irish and Scottish Wise Women, The Cailleach and Her Sisters', given at Woman Spirit Ireland, a Spring Equinox Gathering, *held in Dublin, March 21ˢᵗ, 2015.*

THE PLOUGH-JAGS AND DURGA

From First Furrow to Harvest Hill

If Silbury was designed to enact one phase of an annually repeated drama, this book attempts to restart the same cycle by digging into *midwinter's* hard ground, as a necessary prelude to the birth hill's high summer parturition.

That was also the motive underlying the traditional midwinter English Mumming plays, which were performed in villages throughout England, probably from the time of the Anglo-Saxon invasion, c. A.D. 500, to the 20th Century.

Such rustic shows endured, despite Christian disapproval and 'educated' scorn. So it was that on Christmas day 1852, F.A. Carrington encountered a party of Mummers from Avebury. After performing there, they 'came round the neighbouring villages and, grotesquely dressed, acted their drama from house to house'.[1] Their play dramatised the fight between midwinter's icy darkness and the warm power of returning light, able to bring the start of a new ploughing season. To that extent, the seasonal cycle, as portrayed in Avebury's Neolithic monuments, was given fresh voice by these English yokels.

Harvest Home

Haligmonath, 'Holy month', was the pagan Anglo-Saxon term for September, because then the harvest was nearing completion and so 'they sacrificed to their idols'.[2]

A 19th Century Wiltshire vestige of that long-lost rite, (perhaps corresponding to Silbury's maternal fulfilment), was described by Carrington thus:

'At a Harvest home, care is taken that the last load shall be a light one; and when loaded, it is drawn by the best team, with their bells on. A little boy, with a shirt decorated with ribbons over his other clothes, rides the fore horse. [He is the divine Harvest Child.] On top of the load the rakes etc. are placed. As many as possible of the work people, male and female, ride on the load, with the rest of the party walking on each side. As they go homewards they chant in a monotone: "Ploughed well, sown well; Reaped well, mown well; Carried well, housed well; Nur'a load overdrowd." On the next evening, the master regales them all with a hot supper'.[3]

Mummers and Goddesses

Many Mumming plays from the Eastern counties set the 'winter battle' of the Wiltshire performances alongside the need to woo the primordial goddess, in her manifestations as maiden, bride, mother and crone, by using costume, music, dance, spoken dialogue and song. As in Avebury, the shows were performed at, or close to, the Winter solstice, when the actors briefly re-entered and reactivated a myth-based, goddess-centred 'Other world'. (Writing in the early 8th Century,

Bede associates Yuletide with pagan Anglian worship of the mother goddess at *Mdraniht*, 'Mothers Night'; Old English *modren*, 'maternal'.)[4]

An East Midlands term for the winter Mumming, which emphasises the farm labourers' role, was Plough-Jag.[5] The performance ritually initiated every new arable round by featuring uncanny animals and supernatural characters, combined with those drawn from ordinary life. In essence, such dramas were probably inherited, via an unbroken chain of oral transmission, from the prehistoric inhabitants of Europe, as portrayed in the rock engravings at Bagnola, Italy, and Bohuslan, Sweden.[6]

Divine Ploughmen

Sharing the same roots with the ploughboys of the English folk rites, are Demeter's ever-young ploughman, Triptolemus, son of the mortal king of Eleusis[7], and the former plough god of Ireland, Eochaid Airem; (Irish airem, 'the ploughman'). From the interior of the *sidh*, or fairy hill underworld, Eochaid obtained the art of making oxen work in the yoke, and then became Ireland's original cultivator-in-chief.[8]

Eochaid's double in Wales is Amaethon, the farmer son of the goddess Don.[9] A second legendary Welsh ploughman, Hu Gadarn, was believed to have cut his furrow all the way from India. Of him the 14th Century Welsh poet Iolo Goch declared:

'Hu Gadarn... The emperor of the land and the seas
And the life of the world was he.
After the Deluge he held the strong-beamed plough, active and excellent;
Thus did our lord of stimulating genius.'[10]

Welsh legend asserts that this Hu the Mighty's plough was pulled by large-horned oxen from 'the land of Haf', (Welsh *haf*, 'summer'), or from 'the place where Constantinople now is'. This garbled tale probably reflects the original Neolithic spread of agriculture from the Middle East, and folk rites that bear a strong resemblance to the Lincolnshire plays were known during the early 20th Century in northern Greece, Bulgaria, and throughout the Balkans.[11] Together (and singly) these rites provide something akin to the missing words and actions that may have animated the winter landscape around Silbury.

Should we be amazed by such a link, when the sequence of the seasons, the corresponding agricultural tasks, human life thresholds, and humanity's appeal to supernatural forces were repeatedly woven together by rural communities, every year without fail, between 4000 B.C. and A.D. 1900?

Plough Monday in Lincolnshire

Well into the 20th Century, the Lincolnshire Plough Jags performances, albeit crude and by then fast degenerating, remained fundamentally true to the Neolithic spirit.

The words of the play enacted yearly at Bassingham, Lincolnshire, on Plough Monday – the first Monday after Twelfth Day, (January 6th the old New Year's Day) – were written down in 1823 and form the basis of this study.

Plough Monday was the traditional start of the ploughing season throughout England. Like Luna's crescent moon, the plough's sharp blade, cutting the sod, was the counterpart to Silbury's full moon birth night. (O.E. for Monday was *monan daeg*, or 'moon's day'.)

But before the first sod was turned it was felt necessary to re-enact the annual cycle of the divine powers, in whose footsteps ploughmen and harvesters would then follow. Therefore at Bassingham as elsewhere, the rite began with the dragging of an abnormally big, ribbon-decorated plough around the village. It was pulled neither by horses nor oxen, but by the ploughboys themselves, with up to 20 dragging it along; hence the play's 'Plough Jack or Jag' title. In neighbouring parishes the team members could be called Plough bullocks, or Plough witches. From the outset, therefore, the event retained a prehistoric animal-human ambiguity, together with an evocation of female power, here ascribed to 'witches'.[12]

The close bond between man and beast was also typical of South Wales, where the ploughman's oxen were credited 'with a kind of occult intelligence, like that attributed to bees'. Family events were communicated to the animal, and he was sung to continually when working – 'a set measure, with strains gentle

*A Romano-British bronze plough team, from Piercebridge, Co. Durham;
now in the British Museum.*

and soothing, and a prolonged note or two in each cadence'. The
ploughboy chose words that 'pleased the beast's intelligence... but
mixed with some playful nonsense to gratify his [the animal's]
sense of humour'.[13]

The Lincolnshire play, performed at various places in and
around each parish, including outlying farms, involved a range of
male characters, drawn from the social hierarchy on Earth. They
try to woo a reluctant 'maiden', also played by a man. After rejecting
all these mundane suitors, 'she' is propositioned by The Fool. He
fits into no reputable category, yet is infused with a supernatural
appeal. With all the coolness of early spring weather, at first she
dismisses his advances, before eventually agreeing to wed him. After
their wedding, a baby is born, and the 'mother' eventually reappears,
transformed into her 'old crone' or *Modroniht* 'year's end' state, but
still clutching her high summer offspring. The crone is Mother Earth,
ready to be ploughed again into her former maidenly self, when the
plough will double as the male god, seeking sexual congress with the
prostrate goddess, Earth.

The Plough Jag play encapsulates some of the complexity of the Avebury cycle. Both embody cosmogonic myths that annually attempt to rebuild order from a universe fallen into chaos, with the collapse of vegetation, and withering of the sun's strength. Thus at Normanby, Lincolnshire, Plough Jag troupes from several villagers met around a huge Plough Monday bonfire, on which mistletoe was burnt, and through which the ceremonial plough was dragged, to simulate the arousal of the sun from winter decline. Then opposing troupes held a tug-of-war across the dying embers.[14]

These acts, combined with the ritualistic miming of the first furrow's incision, were intended to help rouse the underworld powers from hibernation, so that the first seeds could be planted in the sacred virgin's ground, there to germinate and sprout.

An Anglo-Saxon 'First Furrow' charm speaks with a similar intent. Just as Silbury was aligned precisely east to west, with its north-to south axis linking Swallowhead to the middle of the inter-causeway moat, so the Anglian charm begins by urging the supplicant to cut four sods, facing the cardinal directions. Milk from every cow in the herd, mixed with parts of every tree, and every well-known herb except burdock, is then to be poured over the four sods. (As with the birth at Silbury, this act was performed at night.)

The charm adds: 'When you drive forth the plough and cut the first furrow say then:

"Hail to thee, earth, mother of men, may you be fruitful under god's protection, filled with the food for the benefit of men"'. Invoking both Goddess and Christian god, the spell is echoed by the coupling of pagan Plough Monday with the Christian Epiphany, celebrated on January 6th.

As in the Stone Age, a good result also depended on sacrifice, so the Saxon recipe continues: 'Then take flour of every kind and have a loaf baked as big as the palm of your hand, and knead it with milk and with holy water, and lay it under the first furrow ...[saying] Erce, Erce, Erce, mother of earth, may the omnipotent Lord grant you fields growing and thriving, flourishing and bountiful, bright shafts of millet crops, and of broad barley crops, and of white wheat crops, and of all the crops of the earth'.[15]

In contrast with these *individually* performed magic acts, the Plough-Jag play was a social event, involving the entire community. Yet the intention was similar; namely to encourage divinity to burst through the surface of the mundane. As in Antiquity, the Plough-Jag rite employed costumed disguise, dance, song *and* magic word. The leading peasant actors *became* the deities they represented, so infusing sacred vitality into a tired, god-forsaken world.

The Play

In common with most Mumming plays, the Bassingham text has three distinct but inter-related parts: The Combat, The Wooing, and The Collection.

The Combat

The combat is between two male protagonists. After an exchange of boasts about their martial prowess they fight with swords till one combatant falls dead. This fight carries connotations of Old Age versus Youth, Winter versus Summer, and Light versus Dark; thus the victorious St George is called *Sun* George in one Cornish play.[16] In this text he complains of being locked underground, prior to his arrival here after midwinter. His opponent, the Dark Knight, alias The Fool, is killed in the fight. As an uncanny 'nobody', everyone has a small share in him. Yet, like the year, he is immortal, and after his death, is always resurrected by a 'Doctor'.

The Wooing

Peter Gelling's study of Sweden's Bronze Age rock carvings shows combat, wooing and solar adoration juxtaposed. These pictograms portray the fertility goddess Freyja, meaning 'Lady', and her husband Freyr.

The 'maiden' of the Lincolnshire plays is also called The Lady. Another resemblance is evident in that Viking Freyja, alias Gefn, or Gefion, 'Giver', brought four giant oxen from the underworld. They were her sons. With them she ploughed Southern Sweden so deeply that part of it floated off to become Danish Zealand.

Like Gefn, and Silbury's sunlit moat, the Bassingham 'Lady' is bedecked with 'gold' jewellery. They bring the underworld, solar power and surface good fortune together, providing they marry. However, Plough-Jag Lady scornfully rejects prosperous suitors from this world, such as judges, farmers or wealthy old men. Instead she chooses the despised Fool, since only he, as a creature of mixed bits and pieces, hints at *totality* and therefore at the supernatural.

In their censored modern form, the birth act is omitted from the scripts, but towards the play's end, The Lady reappears in her role of crone or hag (just as Demeter did in *her* wintry guise). She is synonymous with a 'hag' corn dolly, hung in a winter barn. Here called Old Dame Jane, or Dirty Bet, she carries the August harvest corn child in her arms, but the Fool denies paternity. Only this old mother appreciates the link between cycles. As the repository of memory, she personifies all crops *and* the sacred, renewable Earth. Both she and her corn 'child' are then ploughed into the first furrow to be cut on Plough Monday (*monan daeg* or 'moon day'). Without them, the next year would be barren.

The Collection

When Beelzebub charges in at the end of the play, he raises a club and cooking pan as tools of former combat, doubling as male and female symbols of a divine love match. The pan, (a miniature 'Avebury henge'?), also serves as a collecting box. By dropping in their coins, the onlookers participate in 'seeding' the Lady, and therefore hope to share in her union's prosperous August outcome.

Much of the money gathered was spent on beer for the performers, a drink brewed in Britain since c. 4000 B.C. The remaining coins paid for a candle, lit every night over the ceremonial plough, which was lodged in the parish church till the next Plough Monday. (As at Silbury, the moon was believed vital in promoting the successful harvest at which the Plough Jags aimed. Their solitary candle may have reflected the pagan Anglo-Saxon calendar of 12 lunar months, with a 13th added approximately every three years.)

Following the Silbury – Long Barrow – Avebury cycle, the Plough Jag play integrates seasonal and agrarian concerns with the human life span, in a supernatural drama. The soot-blackened faces of the performers stress their Underworld-Other World origin, while the male-female cross-dressing may evoke an original androgynous state, a feature of many myths. Equally typical of genuine myth is the ease with which the play absorbs contemporary characters, such as a Recruiting Sergeant, without losing the thrust of the archaic plot. (Myth is nourished by history.) The urge to 'do the rounds' of the territory, and the use of hobby horses and musicians, are further signs of the play's ancient credentials.

The Bassingham Plough Play (1823)

Fool: Good Evening Ladies and Gentlemen all this merry time… I've made it bold to call. I hope you will not take it ill what I am going to say. I have some boys and girls drawing on this way. I have some little boys stands at the door. In ribbons they are neatly dressed. For to please you all they do their best. Step in Merrymen all.

> **All:** Good Master and good Mistress as you sit by the fire
> Remember us poor Plough lads
> That runs through muck and mire
> The mire it is so deep
> And we travel far and near
> We will thank you for a Christmas box
> And a mug of your strong beer.

St George: In come I, St George, the man of courage bold. With my broad axe and sword I won a crown of gold. First in a closet I was put, then into a cave was locked, There did I make my sad and greivus mone. How many men have I slew, And runnd the firche dragon thrue; I fought them all courageously, And still got thire victory, England's wright, England's admorration. Now ear I draw my bloody weepon; Ho is the man that doth before me stand? I will cut him down with my courageous hand.

Turkish Knight: Hear comes I the Turkish Knight, Come from the Turkish land to fight; I will fight St George, that man of courage bold, And if his blood is hot, Soon I will make it cold.

…[After an exchange of taunts and boasts they draw weapons and fight. The 'Turk', alias Fool, is stabbed and killed. Then an Old Witch and a Doctor, watched by a King, resurrect the corpse].

Old Witch: Five pounds for a Doctor my husband to cure.

Doctor: I'm the Doctor.

Old Witch: Pray what can you cure?

Doctor: I can cure the itch and the veneral and the gout. All akes within and pains without. You may think I'm mistain. But I can bring this man to life again.

King: Where do you feel his pulse?

Doctor: Where it beats the strongest. {sexual by-play}

King: Is that the strongest part of a man?

Doctor: It's the strongest part about a woman. I've a little bottle in my inside coat trousers waistcoat pocket. He wants a little of my wiff waff just rubbin round 'is tiff taff. This man is not dead. He is only in a trance. Rise up my good man and have a dance.

Lady: [enters singing] In comes the Lady, bright and gay Big fortune makes sweet charms With fingers long and rings upon All made of beaten gold, Good Masters and Good Mistresses I would have you behold.

[Her combination of solar and 'landlady' aspects typifies the poetic amalgams found in early Scandinavian myth and art, and the use made of such compound forms throughout prehistory.]

Husbandman: Here comes the farming man. Upon my principle for to stand I'm come to woo this fair Lady. To gain her love is all my care.

Lady: To gain my love it will not do. You speak too clownish for to woo. Therefore out of my sight be gone. A witty man I'll have or none.

Old Man: Here comes the poor old ancient man. I'll speak for myself as best I can. My old grey hairs they hang so low I'll do the best for myself the best I know. Methinks me sees that shining star bright On you I've fixed my heart's delight.

Lady: Away, away from me begone Do you think I'd marry such a drone? No, I'll have one of high degree And not such a helpless wretch as thee.

Old Man: Kick me Lady out of the room I'll be hanged over the kitchen door.

[In his rejection, Baskerville and Chambers see the displacement and slaying of the old year, in favour of the new.]

Fool: [to Lady] Come write me down the power above that first created a man to love I have a diamond in my eye Where all my joy and comfort lie I'll give you gold I'll give you pearl If you can fancy me my girl Rich costly robes you shall wear If you can fancy me my dear.

Lady: It's not your gold shall me entice Leave off Virtue to follow your advice? I do never intend at all to be at any young man's call.

Fool: Go you away you proud and scornful Dame If you had been true I would have been the same I make no doubt that I can find As handsome a fair one to my mind.

Lady: O, stay Young Man you seem in haste Or are you afraid your time should waste? Let reson rule your roving mind And perhaps in time She'll prove more kind.

Fool: Now all my sorrows is cod and past Joy and comfort I've found at last The girl that used to say me nay She comforts me both night and day.

Beelzebub: In come I, old Beelzebub. On my shoulder I carry a club; In my hand a dripping pan. Don't you think I'm a jolly old man?

[All three dance and sing]: Toorooriadio Tommy's wed tommorio.

Fool: Stop! Stop! Stop! What's all this dancing and jigging about? 'Ere's a tight boy to dance I can dance on a barley chaff riddle. It will neither bend nor break one strand. I am going to ask all you Stickme jacks to me and Lady's wedding and what you like best you must bring on with you. I know what me and my Lady likes. We're going to have barley chaff dumpling buttered with wool and a gallon of ropey ale to it.

[This bizarre choice reflects the hunger often endured in springtime and early summer, prior to the next harvest. Tantalisingly, in this text there is no August-maternity dialogue covering the climax of the agrarian year. Instead, it cuts to introduce a 'year's end' 'old mother', who is still carrying the 'harvest child'.]

Dame Jane: In come poor old Dame Jane Leaping over the meadow. Once I was a blooming girl But now I'm a down old widow [To Fool] Long time have I sought you But now I've found you, Sirrah, come take your Bastard. [She holds her child out to him.]

Fool: [to Lady] Bastard? You Jade, it's none of mine It's not a bit like me I am a valiant hero lately come from the sea. You never see me before now did you? I slew ten thousand men with a seed of mustard

Ten thousand with an old crushed toad. What do you think of that Jane? If you don't be off I'll see you the same.

Betty Besom: [enters with broom] [in the Alkborough, Lincs. text.] Money I want, and money I crave. If you don't give me money, I'll sweep you into your grave.

All: [During the collection they sing]

The mire it is so very deep, the water runs so clear.

Give what you like to our money box. And a mug of your best beer.

Good Master and Mistress You see our Fool is gone

We make it in our business to follow him along. [All exit]

Plough Jag Postscript

The Stamford Mercury noted in January, 1847 that 'the extinction of these heathenish customs from this enlightened and supposedly Christian country has not been achieved altogether', and an 1870 edition of the paper reported that 'the plough boys are still dancing in their uncouth fashion, accompanied by equally rude music'.[17]

What prohibition had failed to eliminate, the following century undermined with increased agricultural mechanisation, urbanisation, secularism and World War I. It was these changes that combined to dismantle the Plough Jag troupes during the 1930s.

Yet in 1932 the equivalent Mummers play, given at Marshfield, Gloucestershire until a flu epidemic demolished the cast in 1880, was revived. In a somewhat corrupt, yet by no means half-hearted manner, it is performed on Boxing Day morning in the village square. There, 'Old Dame Jane' has been converted into 'Saucy Jack'. As the 'bread-winner', he carries several dolls as 'babies' on his back, but complains that they have all died from starvation. He tried his best, but is no mother.

Plough, Harvest and the Church of England

The decline and demise of the Plough Jag rites and English Mummers' performances left a gap in religious experience that coincided with the Church of England's belated willingness to conduct 'Plough Blessing' rites inside parish churches on the Sunday following Plough Monday. This persists in several lower Severn Valley villages, including Moreton Valance. There,

Mr Doug Watts of Moor Farm has adapted a modern plough to replace the horse-drawn implement, previously used in the annual ceremony.

In addition, the Church now conducts Harvest Festival services throughout England and Wales. This was the idea of the vicar of Morwenstow in 1843. He wished to add a sacred dimension to the traditional 'Harvest Home' suppers, given in autumn by farmers for their workers.

The Church officially adopted his initiative in 1863. The full moon nearest the autumn equinox (September 23rd), termed the Harvest Moon, was selected as the time to bring wheat, barley, other vegetables and foodstuffs into the Christian domain. Such produce was afterwards distributed to the 'Poor of the parish'. In the Avebury district these festivals are now held in mid-October, when corn sheaves, canned goods and packets of pasta are laid on the chancel steps. As in the Neolithic, is the sacred to be found again within food?

Silbury and Durga

Contrasting with the restrained mood of these Church Thanksgivings around Silbury, India's October Durga Puja festival is a vibrantly passionate affair. Throughout that country, and especially in Bengal, gigantic, skilfully painted clay images are made of the ten-armed Durga, the great goddess creator-destroyer. Her cult probably originated in the Neolithic Indus Valley civilisation. Ever since then, her annual task has been to demolish, conceive, recreate, and nourish the world, while bringing a harmony of spirit to her devotees, through control of destructive forces, represented by a buffalo demon.

Durga's struggle is re-enacted yearly in a carnival, during which (as with the Avebury group of monuments), the goddess is believed to re-live every stage of a woman's life cycle, from infancy to old age and death.

'Durga Ma', as her followers call her, resembles Silbury as cosmic 'Great Mother' and the personification of Nature, with its extensions into agricultural and urban life. The nine-day festival ends when Durga's image is thrown into the Ganges, or into some other river.

Since 1962 Hindus living in London have celebrated Durga Puja, but not until 2006 did a team of craftsmen from Calcutta come to London to model a 20-feet-high Durga-centred tableau, using Ganges clay and imported cow dung. The festival ended on October 2nd, when at high tide her icon was ritually lowered into the Thames beneath Putney Bridge. There, water flowing from Swallowhead's Kennet (the Thames' major tributary), helped dissolve her image back into slime. In Sangeeta Datta's words: 'As Ma Durga was immersed mid-stream a huge rainbow arched over the sky. Then Durga floated swiftly downstream, clay to clay, and life was a circle once again'.

The word durga, 'inaccessible' or 'difficult to understand' implies that whether richly painted in Bengal, or made of Silbury's grass and moonlit water, the Great Goddess ultimately lies beyond the imagination's grasp. (If she did not include the unimaginable, she would fail to be comprehensive).

Although Silbury, like Durga, is an embodiment of *shakti*, 'the fountain of power', and incorporates the firmament into her moat-body's reflections, the Wiltshire monument is a noble attempt to envisage the unknowable. For example, the Silbury deity probably never revealed, even to her makers, why she gave birth to the Universe.

Consequently, when English Heritage saved Silbury from physical collapse by plugging the miners' cavities with solidified chalk paste, and rendered the central core permanently inaccessible, in order to preserve the Hill's external form, they completed a theologically appropriate act. Thanks to them, 'Silbury-the-Hidden' is now reunited with 'inaccessible' Durga, her inscrutable partner from the eastern end of the Indo-European alliance.

Yet what can reasonably be inferred is that Silbury was designed to re-present a divine birth, as a positive climax to each cycle of death, chaos, re-seeding, germination and replenishment, on behalf of every kind of life form. For millennia illiterate multitudes could to some degree 'read' and identify with the monument's purpose, through the vicissitudes and pleasures of their own lives and with a sense of connectedness to a universal whole. That was what the monumental figure provided. So, under a summer sun they were drawn towards Silbury in time for a full moon's *Gyl Awst*, 'the Vigil of August', and with an urge to participate in the solemn pre-natal ceremony that culminated in the harvest child's delivery, and a shout of joy.

That cry has vanished into the past, yet Silbury is no more 'over' than are the Swallowhead springs and River Kennet 'finished'. Just as their seasonal dive underground is followed by reappearance, so Silbury, arising from a moonlit moat, epitomises rebirth. The great Hill's very shape encourages thoughts of the rotation between death and renewal. Therefore to revolve (if not resolve) the Silbury enigma, amounts to wishing our endangered world many more Happy Returns.

LETTER TO A FRIEND

In attempting to perform the Bassingham village Plough Jag play, using the 1823 script, I saw it as an example of a pre-Socratic, *participatory* mode of understanding; by acting out the relationship between human and superhuman. This involved the eruption into 'ordinary' reality of a divine drama, annually repeated, where the characters and their doings became the model for subsequent human action, in a sequence of farming tasks.

As I tried to say, in such a myth-centred programme, historical particulars are absorbed into the underlying mythic pattern. Thus the names of the Plough Jag characters, such as Turk, Dame Jane, and the Recruiting Sergeant, reflect current or recently experienced events, while giving fresh impetus to the reconstruction of the underlying world of the gods, who are credited with initiating, enjoying and suffering the life-thresholds of birth, marriage, and maternity, while simultaneously being engaged in combat with rivals, who are concerned with death and resurrection.

Given the mythic ambition to achieve a comprehensive synthesis, the above concerns are typically integrated with solar and lunar cycles, both divinely regarded, and with seasonal farming cycles, gathered into a supernatural unity, involving bovine and equine metamorphosis.

Hence the boy Plough Jags are also known as 'Plough-bullocks', as they drag the proto-typical plough around the village.

To dismiss such rituals as 'nothing to do with archaeology', puts my case in a nutshell. Namely that modern Archaeo-*logos* is aligned squarely behind the Socratic position, as disseminated by Plato, which specifically rejected the mythic world view, that, until then had been a universal norm. Against the grain of that outlook, modern archaeology has imposed and institutionalised the Socratic orthodoxies, disseminated by Plato, which specifically rejected the mythic world view, which was subsequently maintained only by 'uneducated' peasants, such as those in Lincolnshire. By contrast, we now have an archaeology of analysis, rather than synthesis, largely concerned with the secular (ignoring the sacred), linear sequential logic and linear abstract time, (as distinct from lived-event).

A dualistic split between mind and matter, (Plato termed the physical the 'Errant Cause'), and hypothesis derived concepts, yielding theory based on 'scientific' proof. Employing instrumental, reductionist means, and these procedures are entirely alien to the myth-based communities of antiquity, which archaeology purports to study. Yet the incongruity of this methodological mismatch appears to go largely unrecognised by those working within the new tradition. Consequently, their admirable achievements in data collection are rarely translated back into mythic terms. Instead, our own technological (logos again) categories and notion of linear progress is superimposed over cyclically performed village time.

The holistic grandeur of archaic existence is denied by our cultural colonialists, and is redefined as error and delusion. By contrast, the word 'history', eagerly taken up by Socratic enthusiasts from a Greek word for 'knowledge', suggests that, at least to historians, that the division of human experience and wisdom into unique unrepeatable moments, is the only valid route to understanding our inheritance. Thus, they argue, that if a version of a mumming play was reworked and performed at a Tudor Court, *this proves* that the material has no connection with antiquity. One might as well claim that because Shakespeare wrote a play titled *A Midsummer Night's Dream*, that he invented the whole idea of fairies. Similarly, rather than claim that

Harvest goddess with sickle, on a Roman mosaic, Brantingham, Yorkshire.

Tudor Court influence cancels the antiquity of village Mumming plays, we might recall that the mythic *hieros gamos*, or 'sacred marriage' between monarch and island-wide 'Landlady', depended on an interaction between village and court, as was demonstrated by the 'May Queens and Kings' of innumerable village performances, with the traffic of ideas and impulses working in both directions, for *the good of society as a whole.*

For our professor colleague then to dismiss Mumming Plays as artificial, 19th Century Merry England phenomena, is to ignore their agricultural basis, and their ancient equivalents in Ireland, (the ploughman god Eochaid Airem) and Hu Gadarn, his equal in Wales, who is said to have ploughed across two continents (with the transmission of agricultural expertise), from Sri Lanka. On passing through Macedonia, he would have encountered a version of the Plough Jag play that is still enacted by villagers, which, (scholars affirm), is remarkably similar to the Lincolnshire form of the play, and is (they add) of undoubted antiquity.

If *the chronological* origins of the Plough Play are to be sought, we might look at the early Neolithic start of agriculture, which in Britain means the fourth millennium B.C. It was probably then that The Maiden-Lady, the Mother, and the Old Hag, were recognised as supernatural embodiments of grain crops, at various stages of their annual development, as they emerged from the ground, and were then harvested. Therefore, throughout the British Isles, the ritually valued corn dolly (idol) is known as maiden, mother *and* hag. She is the Triple Goddess. As the last sheaf cut at the end of harvest she was called either Harvest Mare, or the *Cailleach* (Celtic hag) and hung up through the winter in house or barn, till ploughed back into the land the following spring, to start the next cycle.

The Plough Jag characters can also be seen as the seasonally garlanded female characters who occupy the corners of many Roman mosaics found in Britain. In other words, the theme of the Bassingham play unites the Anglo-Saxon, Celtic, pre-Celtic and Roman periods in Britain. It is the oldest yet most persistently renewed drama that we have.

May I wish you, and my archaeological and historian acquaintances, a happy Plough Monday, which is the first Monday in January, after 12th Night.

Michael Dames

GAEA AND THE GODMANCHESTER TEMPLE

Prior to gravel extraction in an area overlooking the river Great Ouse, an aerial survey of 1990 revealed faint traces of a large ditch and banked trapezoid-shaped enclosure, which was subsequently excavated by English Heritage[1], under the direction of Fachtna McAvoy, with the financial assistance of the gravel digging company. As subsequently reported in *The New Scientist*,[2] he rediscovered what appears to be ancient Europe's most sophisticated astronomical computer, involving both key solar and lunar alignments, with particular emphasis on sunrise at the festivals of Beltaine, (May 1st) and Lammas (August 1st) the solar 'quarter days' marking the beginning of the summer half of the year and the start of harvest respectively.

These and other significant events, including the moon's 19 year cycle, were registered by reading between vertical oak obelisks that were placed around the edge of a trapezoid–shaped bank and ditch structure, nearly seven hectares in size, orientated towards the start of harvest sunrise.

This was an observatory that *worked*, in terms of the community's annual crop growing and harvesting routine, c. 2800 B.C., when the observatory was in use, according to radio-carbon dates obtained from the site. The size of the site was determined by the need for accurate recording of the solar-lunar events, as read between the wooden markers.[3]

Given the recently developed scientific techniques in conducting the modern survey, the prehistoric enterprise under review has been *translated*, whether consciously or not, into secular, scientific terms. The builders are dead. But since sun and moon shine on all alike; and because bread remains our staple food, perhaps we may speak for them, as we eat our next loaf of bread. Our reasonable aim is to incorporate past achievements into a *contemporary* frame of reference.

Therefore we take the Neolithic elemental drama, and strip it bare of the main actors – the deities – and discard the sacred original script altogether, shredding it into the world as machine. But since we are ruled by fact, perhaps we might acknowledge, that prehistoric reality was, as a matter of FACT, a myth-based, recurring event, alien to our notion of wisdom. In 2800 B.C. deities appeared as, and were believed to control, all solar-lunar movements, conducted over what was then regarded as the animated landscape of Mother Earth.[4]

Yet in many respects our culture remains firmly based on a Classical Greco-Roman base involving deities inherited from Antiquity, even when they dare not speak their names.

As mentioned in my introduction to this book, this reticence was evident during my geography degree course, when the name of the presiding goddess Ge[5], or Gaea – founding deity of that discipline – was never mentioned, by the lecturers or other students. Ge, as the very Earth, had become invisible to us urban agnostics, although we daily spoke the word geography, her actual presence, as the founder, inherent in that discipline, was overlooked and implicitly denied. For us, She had shrunk into an invisible abstraction, a two or at most four-letter syllable, inaudible as a living name, either human or divine. Instead, our Mother Earth had become a dehumanised, de-spirited quantity of stuff, stretched out for analysis, into sub-sections on geology, soils, climate, meteorology, geomorphology, and various branches of human geography, like a cadaver, ready for dissection. After performing these subdividing operations, we were then asked to reassemble the bits, into something named Regional Geography; whereupon a shambling Frankenstein-like figure, tried to recreate the initial synthesis, long after he had died of shock on the operating table.

This double procedure was an attempt to balance masculine analytical scientific measurement with the feminine art of reconnection. Even though the latter was offered as a belated postscript, it was considered essential in a course that led to an *arts* degree.

At that time, in the 1950s, C.P. Snow was lamenting the growing division between the 'two cultures' of arts and sciences[6], while Regional Geography's attempt to span the gulf earned derision from both sides. Today, however, the arts have largely imploded, to a stage where archaeologists, who now study pre-scientific culture, do so *only* in 'scientific' terms, supremely confident in their misapplied methodology.

And the world that I so energetically analysed 50 years ago is suffering a perilous climate change, in which Science's 'victory' reads as everyone's defeat.

Some Ge-Gaea Revision
Nearly 60 years after my geography 'finals', it is time for some *Gaia* revision, even though Hesiod's 8th to 7th Century B.C. *Theogony* was

not on our syllabus. Verily, at first Chaos came to be (and Greek chaos was regarded as bounded empty space, devoid of matter, though in Homer's *Iliad*, it is said to be watery).[7]

'But next wide-bosomed Earth, the ever-sure foundation of all. And Earth first bore starry Heaven, equal to herself, to cover her on every side… And she brought forth long hills, graceful haunts of the goddess Nymphs. She bore also the fruitless deep, with his raging swell, Pontus… And afterwards she lay with Heaven, and bore deep-swirling Oceanus… Forthwith she made the element of grey flint, and shaped a great sickle, and put it in her son Cronus' hands. Then, longing for love, at night, Uranus lay about Earth, spreading himself full upon her, whereupon the son lopped off his father's members, and threw them into the sea.'

To this account, the playwright Aeschylus added, 'I give the place of chief honour among the gods, to the first prophet, Earth.' She was the source of the vapours of divine inspiration, and was herself 'oracular'. As the all-producing, all-nourishing Mother, her worship was universal among the Greeks. She had important sanctuaries in most Greek cities, including Athens. Through her child Tartaros, (meaning 'Gloomy Realm'), whom she put underground, she also had control of the underworld. In this arrangement, she was the floor of a primal house, with Uranus providing the roof, and Tartaros the cellar.

By merging the physical fabric of the entire globe into a functioning female form, Ge posed an enigma to us. Her poetic, worldwide, subtleties were best forgotten.

But in ancient Rome she continued to operate as Tellus from Latin *terra*, Earth. There she was also known as Bona Dea, 'The Good Goddess', whose 'women only' festival was held at the start of May. Her temple housed the sacred snakes of the Earth Mother. Sows were sacrificed to her.

Yet another scientific impossibility! For it required geography students to approach the world *subjectively*, from the inside out, as *her* children, conceived by her union with the firmament. Like Hesiod, the Romans were plainly living on a different planet to the one that we inhabited. So too, was the playwright Aeschylus, when he wrote:

Sacred sky longs passionately to pierce the earth
And passion takes hold of Earth to join the marriage
Showers fallen from the bridegroom sky
Make Earth pregnant, and she in turn
Gives birth to sheep and grain.[8]

Yet in considering the 'installation' or 'observatory' built in Godmanchester, Huntingdonshire, c. 2800 B.C., more than 2,000 years before Hesiod was born [*opposite*], we should ask: which world does this 'facility' belong to?[9] Then we shall know whether to call it an observatory or a temple, designed to operate within the context of a sacred universe. If the latter, it is the marriage bed on which celestial bodies mated with the body of a living earth deity, and in so doing made an integrated universe fertile. To help us to decide, pre-historians are in no doubt that a mythic outlook of this kind prevailed throughout Antiquity. Accordingly the Godmanchester monument invites geographers to enjoy a fundamental change of outlook. There, we should *not* ask: 'can we teach them *our* language?' but rather, 'are we willing to learn *theirs*?'

The adjacent Romano-British settlement, which gives the 'chester'[10], ending to the modern name of Godman-chester, is derived from Latin *castra*, 'camp' (and in the 1st Century, the Romans built a pair of camps, flanking their Ermine Street crossing of the river Great Ouse)[11]. This place-name legacy should help us across the threshold, into the 'house of beliefs' that was built right across the ancient world. As for 'Godman', this 'God-man' primal fusion may be only a happy coincidence. The name was spelled in several ways during the Middle Ages.[12] Yet in its present form, it brings a sacred-secular union into focus, and leads humankind towards a divinely granted future.

In order to set their findings securely into the earth, the temple builders created a ditch, scarcely more than half a metre deep, and 1.5 metres (4 ft 11 in) wide, with the excavated material dumped as an internal bank, now largely eroded away. Ditch and bank combined to enclose an elliptical figure, aligned Northeast-Southwest. Around

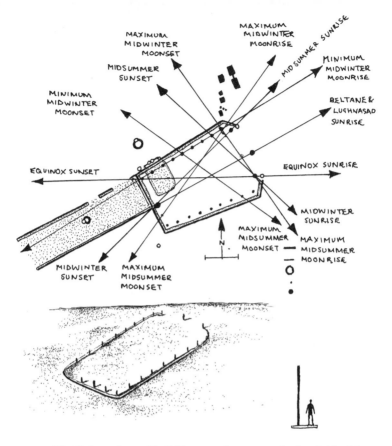

The Godmanchester Neolithic sun and moon temple, Cambridgeshire.
(after David Keys)

the edge of this area, 24 large oak posts, 60 cm in circumference were set.

By reading between pairs of these posts, the risings and settings of the sun, at significant days in the annual calendar, the equinoxes, solstices and quarter days, corresponding to agricultural and pastoral tasks, were recorded. Farming was matched to the arrival and departures of the solar deity's interaction and union with the Earth. Extreme points in the repeated 19-year lunar cycle were also accurately recorded by reading between inter-post lines. Precision was employed and considered vital, in all these religious observations of celestial divine events, upon which human activity was then based.

For the entire community of farmers and herders, life depended on the timely repeated marriages of sun and moon with the receptive Earth. Nor were these celestial events regarded as automatically certain to occur. There was no pre-ordained guarantee that the deities *would* condescend to reappear. Consequently they were carefully monitored, and no doubt encouraged with propitiatory offerings, such as the ox, pieces of whose skull and jaw bones were found flanking the temple entrance, and an ox leg in the temple's ditch, with a possible human sacrifice detected in the form of a dismembered skeleton, buried just outside the extreme west corner of the site. This was a religion of give and take.

So, within the arena, libations probably included water from the nearby river Great Ouse, (whose name derives from Sanskrit *udan*, 'water'), and were probably offered, in a melange of ritually depicted hopes and prayers, combined with festival rejoicing. Such ceremonies of cosmic engagement were believed to ensure that winter's death, and the collapse of vegetation, coupled with the death of human generations, would be followed by springtime renewal, leading through summer's gifts and hazards, to a successful harvest, around which the continuing life of the community swung.

To meet this need, the Beltaine-Lughnasa, sunrise line defined the structure's chief orientation, and main axis, emphasised by two outstandingly massive posts, both 1.2 metres across, aligned on that axis. These coupled the Goddess as *sown* grain, (conceived), with her harvest persona, three months later, (when she gave birth to the First Fruits of her child-crops), on the same solar sunrise alignment. On both occasions, she operated in a polyphonic, multi-layered manner. Her world was based on a practical, grounded poetry, just as in Greece Gaea was depicted in sculpture rising from the ground with the baby Erichthonius, whom she hands to Athena.

At around the time of World War I, the poet and essayist Laura Riding believed that the loss of this organically unified outlook was a disaster. She wrote: 'We cannot get truth because (now) man's consciousness records only difference. He is not a universe-minded being. We can get truth – how things are *as a whole*, only from Women. Man operates through the sense of difference, Women through the sense of unity'. She continues; 'Female time has a rhythmic consistency, with a significance of final repose, while the rhythmic impulse of male time is towards variety and infinite progression'.[12]

To bridge the gap that she had identified, she appealed to an expansion of the 'Inside Sense – the private sphere, the world inside houses, in which men and women jointly live'. She also recommended the transcendent authority of poetry, and saw poets and artists attempting to resolve the male-female division that she had pinpointed, to produce 'a resolution from harsh difference'. While men wrestled with Culture, 'Nature', she concluded, is Woman's referent, and, 'the moral stability of the world, depends on producing a new moral law, formulated from the point of view of the woman.'

Unfortunately this 'law' would appear to defeat the aim of the Male-Female inclusiveness that she was attempting to achieve. Moreover, a female-only Nature would be a sure recipe for the extinction of all life within one generation.

As at Godmanchester, as in Ancient Greece and Rome, the gods have to be allowed access to the marriage bed. But as things stand, this enclosure in Britain offers something that we have forgotten how to enjoy, for we have progressed far beyond what Hesiod defined as the 'Age of Silver' and its matriarchal agriculture. Yet we instinctively return to the monument, (reckoned to be more comprehensive in the range of its observations than is Stonehenge), if only to savour the sense of loss.

Yet though mankind may be imprisoned in the cell of an irrevocably departed pre-historical time, where an obsession with chronology has turned the key and locked the door, surely we can still feel the warmth of sunlight through the bars, and enjoy the moon's night-light.

We might also recall the words of Dorothy Richardson: 'The womanly woman lives all her life in the deep current of eternity... at one with life. Past, present, and future, are together in her, unbroken. She thinks flowingly, with her feelings... with a gift for imaginative sympathy.'[13] Richardson declares: 'my universe = my body + my external world'. Or, as Dora Marsden wrote: 'my universe, minus my body = 0'.[14] Her very existence depended on access to a universe-wide reunion, such as a Godmanchester-like enclosure might provide.

So Gaea of Godmanchester, your wedding *can and should* take place again. For as Martin Heidegger says: 'Man dwells by spanning between the "on the earth" and the "beneath the sky"'. This 'on' and 'beneath' belong together. 'Their interplay is the span that Man traverses at every moment, insofar as *he* is an earthly being.' [15] By honouring this process, Godmanchester's faint traces make a temple that confirms everyone's humanity.

THE RUDSTON GIANT

In Antiquity, all of Britain's earth was regarded in terms of an island-wide living deity, eventually named and figured as Britannia by the Romans. Similarly, the goddess Eriu was considered to be synonymous with the island of Ireland. Yet because of the enormity of the task involved in cultivating these islands, it was often distributed among lesser, yet nevertheless mighty, beings – the giants and giantesses.

The outline of one such 'superman' can still be traced in the cluster of Neolithic cursuses (which are avenues, of uncertain purpose, defined by a pair of parallel ditches and banks) that stretch for miles around the village of Rudston, on the chalk hills of East Yorkshire, to suggest the limbs and torso of a gigantic landscape image.

At Rudston itself, in the village churchyard, rises Britain's tallest standing stone, measuring more than 25 feet tall, and rumoured to be sunk another 15 feet. into the ground. This megalith serves as the erect penis of the reclining giant. Reddish-grey in colour, and weighing many tonnes, it was quarried and transported from Cayton, 12 miles away.

Other Neolithic monuments in the Rudston group include Neolithic long barrows and a henge monument, now named 'Maiden's Grave', and the massive conical Silbury-type mound, Willie Howe. Together they make an assemblage, fit for an annual rite, in which a god (like a proto John Barleycorn), is sacrificed to promote and to personify the next generation of crops. In this context, the cursuses that define his body were intended to contain the primal sacred furrows and seed plantings that initiated and preceded general ploughing of the land beyond his figure, in periods when 'ordinary' farming habitually followed a divine prototype.

Other versions of the same sequence are known across Europe and Asia. For example, the Indian god Purusha willingly sacrifices his 1,000 limbs, in order to make the world. Similarly Romulus slaughtered his twin brother Remus, in order to found Rome, as a divine city named after him.

Anglo-Saxon and Viking settlers of Britain introduced a belief in Ymir, or Mud Seether, as the first gigantic man. He was said to have been created from Chaos, at the beginning of time, when he emerged from melting ice. A hermaphrodite, he gave birth to humanity, and was himself fed by a cow named *Audumla*, (Nourisher), who gave 'rivers of milk'. Eventually, the Sons of Bor slew Ymir, and the Earth was then made from his flesh, mountain crags from his bones, and rivers, lakes, and the entire sea from his blood. His skull was then raised by four dwarves, to make the sky. Around Rudston, the underlying rock is chalk, (which surfaces at nearby Flamborough Head), and is close to the calcium carbonate in our bones. A strongly flowing stream, called 'The Gypsey Race', further animates the district.

This not to suggest that the Neolithic builders believed in Ymir, but rather that successive waves of settlers pragmatically re-used the Neolithic structures to accommodate their own beliefs, resting upon a shared agricultural need; just as the altar of the Anglian Christian church was eventually set up a mere three yards from the great megalith, so surely making a statement of shared need.

The Welsh Giants and Giantesses
A belief in giants and giantesses as active contributors to landscape features was maintained in Wales until at least A.D. 1600. In that year John Dafid Rhys listed no fewer than 50 Welsh giants along with the places where each was thought to dwell. The sites ranged from mountain peaks, such as the top of Cadair Idris, where the Giant Idris had his 'chair'; to Snowden's summit, where Giant Itta or Ritta's body

The Salisbury medieval giant, with Hob Nob, his hobby horse.
Such giants featured widely across England and Wales.

was said to be entombed. Other giants were identified with smaller hilltops, often named *Moel*, Welsh 'bald', implying a gigantic skull reference. Some giants, like Eddern *gawr*, preferred to inhabit caves, while many others were linked either to an ancient summit castle (Dinas), or to a ring enclosure or Caer, of Medieval or earlier date – whether they had founded or taken over these structures. Although some giants, like Crgyn and Bwba, were notably hostile to humanity, others were friendly.

As for their supernatural links, Crbwr Gawr lodged with his three sister giantesses, named Hot Porridge, Warm Porridge, and Morsel of Bread, who echo the Triple Goddess of Antiquity. This trio were killed by Arthur. The Giant Mabon Gawr shares his name with Mabon, son of Modron – a mother and son pair of pagan divinities.

Welsh giants tended to be directly involved in agriculture. Thus Meichiad Gawr kept pigs in Nant Dyffryn Meichiad, while Drewyn Gawr built a caer near Corwen, so that his giantess sweetheart 'could milk her cows there'. Cribwr Gawr's name defines him as 'one who cards wool', and Chwil Gawr's name suggests a Welsh 'circle or rotation', spinning the year around.

Yet perhaps the most substantial and precise gigantic connection to the Neolithic, is provided by Giantesses. Throughout Wales, long barrows of that age are termed *Barclodia y Gawres*, meaning 'the Giantess's Apronful', as if she had personally spilled those carefully heaped rocks onto the ground, which, as a representative of the former goddess of the living and the dead, she had.

The Language Reservoir

If words, in any language, can be relied upon to speak inherited truths; this applies to anthropomorphic terms for our country's physical features. So, on the coast we find English 'headland', cliff 'foot', 'ribbed' sands, and a projecting 'neck' or coastal 'finger'. Inland we may encounter a valley 'bottom' and its 'side', 'flank', and 'shoulder', leading to the 'breast', 'brow' and 'crown' of innumerable hills, that overlook fountain-'heads' of streams, flowing to their river 'mouths'.

To dismiss all such terms as 'merely poetic' is to forget that language is *fundamentally* poetic. It *cries out* to join up and leap between different categories, in order to impart life to the whole. From the surface play of its sound, to the depths of our earth, and of our bones, language can and should be trusted, for it offers us a dwelling-place on this planet.

THE DEEPDALE EFFIGY, BOSTON SPA – TRIBUTE TO THE GODDESS

In a natural, water-worn gulley, running down to Deepdale, which contains a small tributary of the River Ure, in West Yorkshire, the local archaeology group recently discovered an image of an underworld

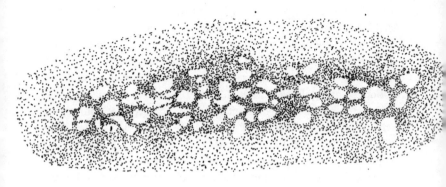

Stone fish in pit no. 8 of the Neolithic Deepdale effigy;
near Boston Spa, Yorkshire.

Mother, apparently credited with the ability to 'give birth' to a variety of animals, depicted as carefully arranged stone settings, in a series of pits, arranged in a sequence, close to her 'pregnant', heavy-breasted pit defined body. Pit 7 was found to contain a depiction of two 'hunting dogs', Pit 6 held what was thought to be a 'cow', Pit 5, a stylised 'wild boar', and Pit 9 held a similarly rubble-created 'fish'. All these 'images' were presumed to have been devised in the Neolithic era, as appeals to the Goddess of fecundity, not least because of the dense scatter of flint, worked in a Neolithic manner, found in the immediate vicinity.

To the excavators, the large central pit defined the head, neck, breast, belly and foot of a seven metre long maternal human image, of a kind known in Western Europe since early Palaeolithic times. The figure was found to contain numerous bones of cows and sheep, and three engraved stones, reminiscent of Neolithic passage grave art. As a 'mother of death and renewal', she continued to be revered, and in the Roman period received tributes of pottery.[1]

So 'this important ceremonial focus for our earliest settled community' has retained something of its significance for at least two millennia, in the director's opinion.[2]

IV.
Topographical Goddesses

SPRING HEADS

Around many a well or spring, memories of a goddess continue to hover. In such spots she spans the threshold between underworld and surface realm to deliver her life–giving waters. Consequently in Britain and Ireland thousands of such animated spots are described as 'holy'. Until c. 1900, every county had at least 40 holy wells, regularly visited for cures and for prophetic guidance. Pembrokeshire recently claimed 236 such holy wells.[1]

The term 'well', employed in this context, follows the Anglo-Saxon usage of *wielle*, meaning both 'a vertically dug water-seeking shaft', and/or 'a natural spring-head'.[2] Our 'holy wells' fit into this broad description.

But the original Goddess naming of these features is often disguised beneath a Christian layer, involving reattribution to a female saint. For example in Derbyshire there are five dedications to St Anne, (a 'saint' derived from the pagan deities Anu and Innana, controller of cosmic waters and childbirth), along with two for St Helen, plus eight examples of Lady's Well.[3] In many areas, including London, wells are named after Bride, alias St Brigit, the Irish fire and water deity.[4]

In England, pilgrims were advised to approach a holy well from the east, and at dusk, in keeping with the 'threshold' nature of the event. Then they were expected to circulate three times around the site, in a sun-wise direction, and in total silence. After that, they offered their prayers, drank the spring water, stored some in a bottle, and made an offering of a bent pin, thereby signing a contract between surface reality and a supernaturally regarded underworld. Then they normally lay on the ground, close to the outflow, in an 'incubation' slumber.[5]

Among the wide range of maladies for which cures were sought; including children's whooping cough and rickets, and infertility in women, eye malfunctions are also frequently mentioned. This last emphasis reflects the regard for the well as the divinity's watery 'eye', doubled in strength after sunset, when the 'eye in the sky' lay beneath the horizon.[6] At night, an eye-to-eye encounter with the springhead

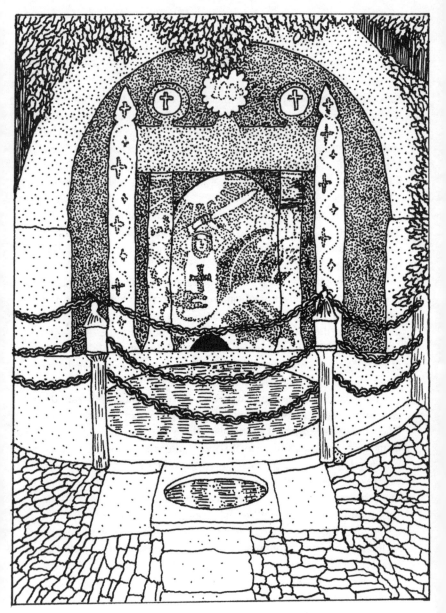

The Yew Tree Well, Tissington, Derbyshire. Dressed in May, 2006.

gave access to the well's illuminated thoughts and intentions. Likewise, in Hinduism, *dasein*, 'eye contact' with the deity is highly valued.

After that exchange, the supplicant would fix the rag that had been bound around the damaged limb, or other troubled part of the body, onto a nearby tree. In time, when this cloth disintegrated and fell to the ground, their cure might take effect.[7] However, some tidy-minded visitors now object to these and other visible tokens of belief, and systematically remove them, so that their own encounter with a supposedly virgin Nature can appear untouched. While, prior to emergence from its underground tunnel, the spring water is undoubtedly pure, by contrast at the surface it has entertained countless generations of supplicants and lovers, in a shared, not to say crowded world. The springhead offers a *point of exchange* between Nature and Culture, in all its many formats.

It was customary for all visitors to leave the spring before sunrise. In order to sustain empathy with the underworld's state, the pilgrims conducted and completed their beneficial encounter with the spring and its supernatural spirits, entirely by moonlight, or in total darkness.[8]

Looking back, the holy wells of Britain seem to have plaited together strands of belief and ritual from remote prehistory, with later Classical, pagan, Anglo-Saxon, Viking, and Christian cultures. There is room for everyone, of whatever race or creed around a holy well, and Christianity's attempts to impose an intolerant mono-cultural reading onto springs, have conspicuously failed.

Today's rag hangers incidentally serve to emphasise the Germanic importance of the well-head tree, as a sacred component in the ensemble. Specifically, via England's adoption of continental myths, the ash tree, named *Yggdrasill*, holds the universe together. It has sky-high branches, and roots penetrating far below the subsoil, deep into the world's rocky interior, while its trunk props up the entire cosmos.[9]

Coincidentally, a self-planted ash tree has now established itself, two yards from the Swallowhead springs at Avebury. As at Swallowhead, Yggdrasill's roots were fed and sustained by a sacred spring named *Mimir*, a word cognate with English 'memory'. Mimir was regarded as the supreme source of wisdom, and gave the tree its alternative title, *mimameidre*, 'the tree of Mimir'.[10] Yggdrasill, like the Swallowhead ash tree, can be seen as a vertical fountain of foliage, and another version of the 'childbirth' metaphorically enacted by the emerging waters. Accordingly, in Nordic myth, the first man was born from a tree, and was named Askr, 'Ash'.[11] In Hindu terms, Yggdrasill served as the male *lingam* to the well's female organ, vividly portrayed by another spring, named Hvergelmir, which also fed the Great Tree's roots. Hvergelmir literally means 'resounding kettle or cauldron', from which rivers were said to flow. In Hindu terms it provides the *yoni*, to pair with the great ash as *lingam*.[12]

How then, can we forget that the Swallowhead springs, surrounded as they are by a major group of Neolithic monuments, call us to remember past achievements, as Mimir's name and wisdom requires, and that Swallowhead is also the birthplace of a major tributary of the River Thames, or as some claim, the originally acknowledged source of the Thames, namely the Kennet or Cunnit, so called after the female birth-giving orifice.

Until recently, I have tended to underestimate the English contribution to the appreciation of our greatest Neolithic monuments, so it is time to correct this oversight, and to welcome also the third 'tree sustaining spring', that the Norsemen called Urdr, or 'Fate'. Urdr resembles English *weird*,[13] recalling the ineluctable processes of Nature, to which even the Gods must submit. Thus at Swallowhead, the role of delivering the 'head' of new life, along with an entire river, is accompanied by 'swallow' – the removal of the dead and dying, helped by the corpse devouring snake, Nordic Nidhoggr, who lives in the world tree. Thus the Swallowhead springs offer a thought-provoking display of intermingled living truths, which apply equally to the deer, squirrel, hawk and eagle, which are said to live in Yggdrasill's branches.[14]

Links between the springs and the Great Tree are strengthened above ground by its habit of dripping *aurr*, 'a clear life enhancing liquid', and also by secreting *Heidre*, 'the sap of life'.[15] Yet the same tree is also the 'gallows tree' upon which Odin hung himself. Meanwhile, the three giant goddesses, known as the Norns, sprinkle water from Urdr's well over the trunk of the World Ash, which then turns white, as does the milk-white chalky water, emerging from the Swallowhead springs. The three Norns were each said to embody one phase of the moon's cycle, either as crescent, full, or waning moon. As a trio, they match the Celtic Triple Goddess. The Norns' individual names translate as 'Past, Present and Future'.[16] They can measure the lunar aspect of the nearby Silbury birth-event, while as 'The Fates' they determine the length of every human life-span.

Swallowhead is flanked on the south side by one of Britain's most spectacular chambered long barrows, which is connected to Silbury, according to 18th Century belief, by an underground tunnel. Indeed, as Colonel Drax noted in his letter of 1770, Silbury's massive 130-feet high 'pregnant womb' is precisely centred over a vertical shaft that he was unable to fully fathom.[17] The 'shaft' may have been a natural pipe, worn in the chalk by rainwater acidity, around which the great Silbury mound may have been deliberately centred, to suggest that Silbury's entire upwelling was an underworld gift. (And can a single ear of corn be found that *does not* depend on its underground roots?)

Rising so close to Silbury, Swallowhead appears to balance light and dark, life and death, pregnancy and adulthood, and provides a boundary between visible and hidden worlds.

In seeking the original identities of England's spring heads, hints are provided from across the entire Indo-European spectrum, especially when associated with the start of agriculture. For instance, the Babylonian goddess named Bau, who dwelt in the primordial spring that was overhung by *Kiskanu*, a tree, has a range of functions. Bau was the goddess of plenty, responsible for flocks and crops; her son, Ea, introduced arts and crafts, while his sister Nanshe shared his functions, and was also the goddess of springs. At Lagash a procession of boats escorted the sacred barge in which Nanshe rode.[18]

Whatever their names, we may reasonably suppose that the divine Neolithic spirits of Swallowhead embodied roles similar to those of Bau, Ea, and Nanshe. Finds of Mesolithic artefacts *pre-date* the start of farming, around Swallowhead, implying that its value was recognised throughout the subsequent Neolithic era, beginning c. 5000 B.C.[19] Its virtue continued to be acknowledged during the Bronze and Iron Ages, and a considerable amount of Roman material was excavated from the spring's immediate vicinity by boys from Marlborough College, while as recently as the 18th and 19th centuries A.D., people continued to revere the water, and annually took it to the top of Silbury.[20] According to William Stukeley, local people also revered the Fools Watercress, *apium nodiflorum*, found growing around the spring. (In Western England that plant is still incorporated into meat pies.)

Swallowhead was never claimed by a Christian saint, while signs of today's pagan devotion can be seen in the variety of gifts donated there – statuettes, pots, flowers, fruit, and poems, all left on the white chalk rock around the fount's dark mouth; or hung, amongst bits of cloth, and messages in several languages, which are left suspended from the nearby trees. Even aloof archaeologists are now, if belatedly, accepting the springs' eloquent revived appeal.

Fundamental to spring-head attitudes and behaviour was, and is, the notion of 'Give and Take'. Recognition is delivered, along with gifts and other signs of veneration, and in return, the pure water flows forth, straight from the underworld's vagina.

From Swallowhead, the spring-water trickles for twenty yards to join the Winterbourne stream, to become the River Kennet, which provides more than half the Thames' water during the summer months.

Prior to the 20th Century's nation-wide installation of piped water supply, every spring offered a rare life-saving gift, apparently provided by the well-head persona. The sheer generosity of a stream, gushing unaided onto the surface, can affect the emotions, even among contemporary mains-supplied onlookers. Rather than dismissing this continuing attraction as a misplaced, old fashioned superstition, we might see it as a valid attempt to join individual awareness to the great wheel of life, at locations that invite the perception of unity.

Indeed, if 'connection to the mains' *disconnects* us from the landscape drama playing at our feet, in which we have been given a walk-on part, it would amount to a harmful deprivation.

Every holy well in Britain has its own traditions, developed and modified over millennia; of these a few examples are offered below.

In Derbyshire and the north Midland counties, the practice of annual Easter-time well dressing is still flourishing. This involves erecting a large, light-weight wooden screen around the well-head, onto which a layer of wet clay is smeared. This provides the base for inserting flower heads, arranged in intricate pictorial designs, often illustrating biblical themes. The Derbyshire village of Tissington has displays around each of its six well-heads, with the blooms lasting for a week after Ascension Day.[22]

Suspicion of holy wells has long been recorded. St Anselm issued an edict in 1102 A.D. 'that there should be no adoration of fountains without a bishop's permission'.[23] In 1540 Sir William Bassett boasted in a letter to Lord Cromwell, that he had removed and defaced two images of St Anne from St Anne's Well, Buxton – one of which had been 'miraculously found' in her well. He reported he had thrown away crutches, shirts and shifts that had been offered there, 'to entice and corrupt the ignorant'. He then locked and sealed the well.[24] Morris Dancing and a well-head chapel had previously brought Christian and pagan responses together; her re-opened well now continues to be much visited by both Christians and non-Christians.

Similarly, and despite Puritan hostility, 49 carts and wagons passed through Baslow, Derbyshire, blaring trombones and carrying flags, on their way to the Pilsley Well dressing. Open House was kept by local people for the occasion, and there were stalls selling nuts and gingerbread.[25]

Mermaids frequented some Derbyshire wells. A mermaid who lived under Kinder Scout came to bathe in the well at Hayfield.[26]

It was believed that any man who saw her there would become immortal. At nearby Chapel-en-le Frith, a mermaid's reflection can be seen in Mermaid's Pool at 12 o'clock on Easter Sunday.[27] Likewise at Rostherne Mere, Cheshire, a mermaid rises on Easter Day, ringing a bell, before disappearing down a tunnel that was said to lead to the Irish Sea.[28]

For the Norse gods, there is Woden's Well at Wanswell, Gloucestershire, and Thor's Well at Burnsall, Yorkshire, while Chester has an altar dedicated to *Nymphis et Fontibus*,[29] which directly credits the Nymphs for that Roman city's water supply. Prehistoric bronze spoons have been found in pairs at some wells. In Cumbria, at the Eden Hall Giants' Cave well, 'a vast concourse of young people gathered on the third Sunday in May, to drink sugar sweetened spring water, afterwards celebrating in a public house.'[30]

In Cornwall, which has more than 200 holy wells, an old woman presided at the Gulval outflow. She relayed the well's secrets, and told the fortunes of pilgrims, based on whether the spring bubbled or was discoloured. This priestess died in 1748.[31] At Menacuddle Well, St Austell, sick children were brought to bathe,[32] while at St Nun's Well, at Penlynt, pins were offered to 'the Pixies' for their goodwill and help in agriculture'.[33] Attempts were made to cure the insane, by ducking them in St Clear's Well, Liskeard.[34] Crowds of people visited Our Lady of Nantwell, near St Columb Major, on Palm Sunday, each carrying a palm, which they threw into the spring. If their palm sank, they would die within that year.[35] Meanwhile at St Catherine's Well, Milton Abbey, Dorset, young women habitually begged that saint for 'A husband, A handsome one, A rich one, A nice one, and soon.'[36]

In Wales, St Winifred's Well in Denbighshire is much visited, while St Illtyd's Well near Swansea, instead of water, is said to flow with *milk*, from the breast of the Mother Goddess.[37] Butter has been made from her outpourings. In his survey of Welsh holy wells, Francis Jones reports frequent sightings of the White Lady, *Y Ladi Wen*, hovering around these sources.[38] Perhaps she embodies the water-born memory of a long-forgotten deity, seen returning to her birthplace.

How ironic that while Leary and Field's approach to Silbury has removed for them any trace of the Great Goddess' divine imagery or influence from that edifice, yet they simultaneously welcome the restoration of sanctity to the spring that almost certainly helped to inspire that monument's construction.[39]

Due to a recent fall in the level of the water table, the Swallowhead springs are now only intermittent in its flow. In summer it is now bone dry, yet faintly it still cries out to offer us all a fresh invitation to rediscover our whole land as a living entity, season after season, and thereby rediscover the key to understanding the meaning of all the Avebury monuments.

RIVER GODDESSES

'The valley spirit never dies. It is the woman, primal mother'. So wrote Lao Tsu in his *Tao Te Ching*.

Rivers don't care about people. From oozy mountain bog to coastal mud flats, they answer instead to *gravity*, rather than to humankind. Yet people have always needed *them*, both for the practical affairs of life, and as an important element in patterns of worship, from birth to death, and from source to sea. Consequently our rivers have often been seen as embodiments of the Goddess, in her watery form. And as ever-running highways of devotion, many of our rivers retain emphatically supernatural names, even while draining disenchanted territories.

In Britain there are four rivers that are still named Dee, a term derived from Celtic *Deua,* cognate with Latin *Deva;* words meaning both 'Goddess', and 'The Holy One'. These Dees include the Aberdeenshire Dee; the Dee of Dumfries and Galloway, southwest Scotland; the Cumbrian Dee; and the Dee of north Wales, which runs to Chester, a Roman city, originally named Deva. In Ireland, the Dee river runs from a spring in County Cavan through to Dundalk Bay.

In the 7th Century the Welsh monk, Gildas, waved a premature goodbye to the tradition of river worship, which was probably as old as the human presence in Britain. Yet despite the negative demand of his calling, he did so, with some reluctance, in writing:

'To water it, the island [of Britain] has clear fountains, whose constant flow drives before it pebbles white as snow, [The White Lady, on the move], and brilliant rivers, that guarantee sweet sleep for those who lie on their banks'.[1]

Yet should the sleeper dream again of the sanctity of these waters, he goes on to warn: 'I shall not name the mountains, hills and rivers, once so pernicious, now useful to human needs, upon which, in those days a blind people heaped divine honours.' At a stroke, he claimed to have converted adoration to mere utility, so recommending a detached exploitation of nature's resources.

But a loving relationship with rivers was harder to eradicate than Gildas supposed, not least because of the account of creation in Genesis, in which God created dry land from the waters, which served in an amniotic manner, first protecting and then delivering the entire world. Having established the Earthly Paradise, God then equipped it with four primary rivers flowing north, south, east and west. (Attempting the Biblical four, Pumlumon in mid-Wales manages only three river sources, those of the Severn, the Wye, and the Rheidol.) There is often a gap between ideal and real.

Of the four elements, water is regarded, in almost every mythology, as the *most female*. Water acts as a receptive, if unpredictable, life-producing, and welcoming solvent, shared by all life forms.

This indispensable value encourages the river Dees, as serious divinities, to flow on, along with other British rivers that retain traces of a goddess connection. For example the Lleyn peninsula's Dwyfor and Dwyfach streams, pair 'big' and 'little' 'goddess' names, as they run through districts where prehistoric monuments abound.

In his standard work on English river names, Eilert Ekwall has no doubt that Anglesey's Braint, derived from the Welsh word *braint*, 'privilege, highness', alludes to its sacred inheritance, as does London's river Brent, a name which he identified as deriving from the Goddess Brigantia, herself a form of O.Ir. Brigid. Brent, he states, 'is a clear instance of river-names pointing to river worship, for it means the exalted, holy river', while to him Lancashire's river Lune – 'the health-giver' – suggests 'the abode of a divinity with healing powers'.[2]

Other English rivers such as the Scutterskelfe, running near Rudby, Yorkshire, derive their names from O.N. *skradva,* 'to talk, to chatter'. It can be associated with innumerable 'babbling brooks', all hinting at the gift of speech, delivered in a dialect hard to understand, yet believed to offer audible evidence of supernatural intelligence.

Yet Christian pressure has concealed the divine attributes of many of our streams. So the goddess Belisama, inherited from a Gaulish form of the great Minerva, as 'The Shining One', has now become the Lancashire Ribble (O.E. for 'tearing'). Meanwhile, our neutral sounding Avon, simply meaning 'river', (of which there are *six* examples), were each formerly called *Abona,* with the *-ona* ending, widely accepted as a clear sign of feminine sanctity.

The infant River Severn, Britain's longest river, seen above Blaen Hafren, mid-Wales.

Other vague descriptions of rivers, such as the Dark One, (the Thames), have been superimposed over more explicit earlier terms, after goddess names became taboo. So the Cumbrian Dee has now had the alternative name of Dent imposed upon it.

Similarly the abstract sounding name Severn, 180 miles long, is directly derived from that river as a water goddess, Sabrina, who is named on the maps of the Classical author Ptolemy, writing in the 2nd Century A.D.

Sabrina flowed on, to feature in Geoffrey of Monmouth's 12th Century account of a struggle between the eastern and western powers of the country. In London, he says, the king kept his mistress Estrildis in a cellar beneath his palace. There she gave birth to a daughter, Sabrina, who was then abducted, and drowned in the Severn, by his jealous wife Gwendolen, whereupon Sabrina became that river's perpetual life-force.

She brought her mother's eastern sunrise power, together with her cellar-dwelling experience of the underworld, as her endowment. Her own name is said to derive from Sanskrit *Sabhar-dhuk*, 'milk'. The fertility of the pastures through which she flows was believed to be enhanced in the early 18th Century, by the custom of sprinkling flowers into her river, at sheep shearing time, as described by the poet, John Dyer. This custom was followed in Shropshire between the Breidden and Wrekin hills.

Pale lilies, roses, violets and pinks,
Mixed with the leaves of burnet, mint, and thyme.
Pleased with the honour due, Sabrina, guardian of the
crystal flood

Shall bless our cares when she by moonlight clear,
Skims o'er the dales and eyes our sleeping folds.
No taint worm shall infect the yeaning herds.[3]

So wrote the 17th Century poet, John Milton, in celebration of Sabrina's continuing life in the river, from which, he says, she sometimes rose, along with her attendant water nymphs, to sing and dance on its banks. Between 1770 and 1995, the inhabitants of the Severn-side city of

Worcester drank Sabrina daily. She issued from their taps, because their water supply was drawn directly from the river.

Having risen at nearly 2,000 feet above sea level, on the mid Welsh Pumlumon ('five beacons'), she eventually runs into the tidal Bristol Channel, from where the ocean's moon driven tides surge upstream in the famous Bore wave to Gloucester, bringing with it millions of elvers or young eels.

Sabrina epitomises Nature in the west Midlands, for, as a valley-carving agent, she has largely created the land through which she flows. Illustrations to Michael Drayton's long topographical poem *Poly-Olbion,* published c. 1620, show every stream and river in Britain inhabited by a water deity, typically female – clear evidence of previous river goddess veneration throughout Britain.

The 16th Century Protestant Reformation, (combined with a Renaissance emphasis on the individual), produced another vision of natural synthesis. As a result of the individual being granted direct communication with godhead, as Protestants believe, the human body became a plausible microcosm of the entire divinely regarded universe. Consequently, the dendritic pattern, made by rills and brooks feeding into major rivers, came to be compared, and equated with, the array of veins, capillaries, and arteries that make up the blood's circulation system within the human body. From this viewpoint, each person *became* a miniature Britain, with our rivers literally recognised as our main arteries.

After the development of trading links with India, yet another perspective on the sanctity of rivers became available in Britain. News of the Ganges, there regarded as Divinity in liquid form, filtered back to Britain. She was, and to Hindus still is, the epitome of *Sakti,* female energy, and is utterly auspicious. India recognises at least seven other holy rivers, believed to supply the whole country with sacred waters. British contacts with Egypt and West African nations also helped to bring home the realisation that our own divine river legacy was part of a world-wide endowment, which included the biblical four rivers running from Paradise.

Although rivers always flow in the same direction, in human terms they generate a two-way response. This has involved the ritual offering of precious gifts, including, during martial periods, exquisite pieces of military gear, such as the famous Iron Age Battersea shield, one of innumerable prehistoric offerings made to the river Thames. Similarly rich endowments have been found in Lincolnshire's river Witham, and at Flag Fen, where in 1982 Francis Pryor recovered 330 ritually deposited bronze items.

Human sacrifice offered to our rivers may be suggested by the prevalent folk belief, that rivers claim and *need* at least one drowned victim a year, to assuage their own hunger, and to secure their relationship with humanity.

For example, the river Dart in Devon, the Tweed in Scotland, and the Aberdeenshire Dee, are among many waterways said to be hungry for human life, according to our folk rhymes and local records:

'Dart, Dart, cruel Dart,
Every year thou claim'st a heart,'[4]

Similarly, the river Till boasts to the river Tweed: 'ye droon yin man, I droon twa'; while Bloodthirsty Dee each year needs three. (The Tweed's 'fairies' had salt offered to them at the start of the fishing season, after the vicar of Norham-on-Tweed had blessed the nets and boats.)

These annual requirements may be seen as a continuation of the prehistoric tradition of ritual sacrifice of honoured individuals, such as Lindow Man, drowned in Cheshire. Whether accidental or deliberate, river deaths evoke the old pattern, and to detect the position of a drowned body, a loaf of bread was sometimes floated on the stream, in the belief that it would come to rest over the submerged corpse.

River deities can adopt animal form. Versions of the divine cow of the Indian Ganges, and the Bull god of Mesopotamian waters were known in the British Isles. Thus Ireland's river Boyne or *boann,* is synonymous with the sacred 'White Cow', which features in Irish myth. In Wales, her equal stands overlooking the source of the river Severn, as *Buwch Wen a'r Llo,* a pair of white quartz rocks, known in English as the White Cow and Calf.

The beaver (Welsh *afanc*), also plays an active role in legendary river affairs, as does the horse. In Scotland he is termed a kelpie, and is noted for inviting unwary wanderers to ride on his back, whereupon he gallops with them into the nearest river.

Some British Water Spirits
Since the many uncanny beings known in Scotland are invested with a measure of supernatural power, and often demonstrated a shape-shifting ability, they may be seen as intensely local modes of a broader spiritual presence:

The *Kelpie*, or water-horse, could assume the form of a beautiful, naked, bathing nymph.

The *Bean Nighe*, or Washer at the Ford, wore green and had webbed feet. She foretold deaths while washing clothes. Her ford was the Scottish equivalent of the Indian *tirtha*, a life-death crossing point. The river Clyde (from Celtic *clouta*, 'to wash' or 'cleansing river') is named after her.

The *Caoineag*, a wailing hag who lived under waterfalls, was known as The Weeper, and she also foretold death. The waterfall provided her with 'tears'.

Crodh Mara, the Cattle of the Sea, are the herds of Highland fairy water cattle that emerged from the rivers.

Each-Uisge was a supernatural water horse of Highland lochs and the sea.

Fachan – a Highland water sprite, with one leg, and one hand.

Gentle Annis – a water spirit who lives in the Cromarty Firth, North East Scotland.

Loireag – a water fairy of the Hebrides, patroness of spinning.

Noggle – a mischievous Shetland water-horse which led the Vikings to adopt the water horse idea.

Selkies, in Scottish, Irish and Faroe folklore, are seal creatures – which take human form on land.

Urisk – a half goat, half human that haunted pools and waterfalls.

Water wraiths – female Water Spirits with withered faces, and dressed in green. They could drag people to the depths.

Cailleach Bheur – the most feared Scottish hag, who stood for Nature's destructive aspect, in particular.

The Carlin is her English counterpart, an old hag whose name is of

Nordic origin. She is said to inhabit, and has given her name to, many of the mountain rills in Cumbria.

The *Nix* or plural *Nixie*, is a shapeshifting water spirit with Germanic roots. They share the rivers with the *Neckan* river spirit.

The *Phynnodderees*, are naked, hairy water fairies of the Isle of Man.

The bat-like *Water-Leapers* live off the Welsh coasts.

Yorkshire has the *Gindylow* water sprite.

Lancashire's *Jenny Green Teeth* drowns many a river bank loiterer, as does *Raw-head* and *Bloody Bones* in the same county.

Peg O'Nell now claims the Ribble as her own, while *Peg Fowler*, a Lancashire river deity, operates in the Tees[5].

Throughout the country, under several local names, *Will o' the Wisp*, plays his light over marshes.

The Cornish *Cyhiraeth*, 'the Washer at the Ford, a withered portent of death', runs in streams towards the Shoney *(Seonaidh),* the Cornish Goddess of the sea and a Celtic water spirit.

The *Knucker* (from O.E. *Nicor*) is an English swamp dragon.

Individually and as a group, this array of river-related beings reminded people that their landscape was shared with other creatures, real and imagined, in a 'poetic of habitat' – an insular form of the Classical appreciation of *genius loci,* 'spirit of place'.

A BEVY OF 'SILBURIES'

In Ge-omorphology, as modern studies of our surface landforms are termed, Ge is acknowledged everywhere on the surface fabric of our island's landscape.

Multiple Epiphanies

Silbury is unique. From the outset the monument may have attracted island-wide attention, as indicated by the Neolithic stone axes found in the vicinity, some made in Cornwall, North Wales and Cumbria.[1] Yet because Silbury embodied a commonly held perception of the Earth as an active, harvest-giving deity, that belief found physical expression in many other places.

Throughout Britain and Ireland people were drawn to evocative features in the natural landscape, such as isolated hills that were seen to embody the same procreative theme. Alternatively, as at Silbury, they *constructed* a 'pregnant' mound, *or* artificially modified an existing hill, to enhance its suitability for First Fruits ritual performances.

Silbury invites us to renew an acquaintance with many other former 'harvest hill' ritual sites, scattered throughout the land. Some other examples of natural, modified and constructed hills will be given in this chapter, since they each throw additional light on the meaning of Silbury and demonstrate the persistence of Silbury-type behaviour and belief.

Paviland

In his 1998 study of the Palaeolithic goddess figurines, carved 22,000 B.C., that were found in Paviland cave, Gower, Aldhouse-Green remarked upon 'the numinous quality of the site'.

'If viewed at low tide, from below, on the plain, now drowned by the Bristol Channel,' one can share in our forbears' pleasure.[2] Seen from there, a 50 metre-tall truncated cone of white limestone rears skywards as 'a natural landmark... sacred or imbued with mythic importance, [an attitude] widespread in the ancient and pre-industrial world'.[3]

Carved entirely by Nature, the cone head displays 'eyes', set beneath a beetling brow, and a gaping mouth, with the lower cave serving as her uterus. Paviland offers an enduring example of topographical imagery, pre-dating Silbury Hill in its Goddess-orientated appearance and usage. The ceremonially buried dead, both human and mammoth, along with Goddess images of carved bone, were laid in her 'womb' 17,000 years before agriculture came to Britain.

Llyn Y Fan Fach

Continuity between Neolithic and early modern times is evident in 'Celtic' Western Britain. There the population, which retains traces of a *pre*-Celtic inheritance, were, until recently, well disposed towards

'Harvest Lady' legends, centred on memorable natural features. Such was the case at Llyn y Fan Fach, a small lake set below the Welsh Brecon Beacons.

Oral accounts from 1861 tell of a widow who sent her grown-up and only son to graze her cattle near that lake.[4] There he saw a beautiful lady sitting on the smooth water, which she used as a mirror, while combing her long ringlets. He offered her some of his barley bread, and tried to touch her. Eluding him, she cried: 'Hard baked is thy bread. It is not so easy to catch me!' Then she dived, disappearing under the water.

When he came home, his mother told him to try unbaked bread. The next morning, before sunrise, he returned to the lake with raw dough to give the phantom. On reappearing she rejected it, saying: 'Unbaked is thy bread. I will not have thee', and submerged again.

Throughout a third day, now carrying the *slightly* baked bread that his mother had cooked for him, he waited by the *Llyn* from dawn till dusk. Eventually he saw several cows moving across the water, followed by the lady. She approached, took his hand, and agreed to be his bride; but would break off their marriage if she received three causeless blows from him.

By the following spring, to his chagrin he *had* struck her three times, albeit only slightly. The blows fell at a christening, a wedding, and a funeral; whereupon the lady called the cattle to join her, and both she and the herd sank back beneath the lake.

'The Lady' was the goddess of the annual cycle, owner of the first domesticated cattle, and the world's first baker. Having overseen the three mortal thresholds of birth, marriage and death, she and her animals retired to the netherworld.

Yet future generations were eager to witness a repeat performance for themselves. Accordingly, as an old woman from the nearby village of Myddfai recalled in 1881, in her youth 'thousands and thousands of people visited the lake on the first Sunday or Monday in August' when 'a commotion took place' on its waters. They 'boiled'; and this was a sign that the lady and her oxen were about to reappear.[5] Both the calendar date and the 'boiling' (Welsh *berwi*) are reminders of Silbury and *its* posset of milk, which by repute seethed as that Hill arose.

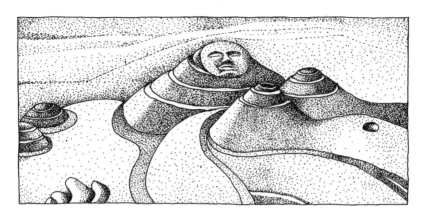

The Cramlington effigy in Northumberland, created by C. Jencks in 2010 from coal mine debris.

In 1858 it was reported that both sexes visited Llyn Fan y Fach on the night before the first Sunday in August. This Sunday represents a Christian variant on the pagan Welsh *Gwyl Awst*, 'August Vigil' or night watch, held at full moon nearest to the August 1st quarter day. This was when the Goddess was in labour, as reflected by the 'moon-on water' Silbury birth event.

At Llyn y Fan Vach the pilgrims also expected to 'see a host of fairies', The Lady's lake-born offspring. They were supposed to emerge from rocky crevices at the lake edge, and skim their 'mazy dances across its surface'.[6] Beneath them swam eels, plentiful there in August, according to a Llandovery fishwife.[7]

Other people spoke of an 'ancient town' under the water that could only be glimpsed at sunrise on August 1st. (Apparently the Other World could take an urban form, comparable to Roman Caerwent in Gwent, built for the Silures tribe in the 1st Century and still inhabited today.) A red sandstone mountain wall, over 220 metres high, rises above the tiny lake. Across its sparkling water the August Lady came. Dressed in white, her long yellow hair gilded by sunbeams, she sometimes arrived paddling a golden boat with a golden oar.[8]

After she and her cattle had disappeared again, it was believed that her son guarded the herd on the adjoining Black Mountains, where the rivers *Twrch* ('Boar'), *Gwys Fach* and *Gwys Fawr*, ('Little and Big Sow'), meander southwards through those rough pastures. As at Silbury, swine and goddess are often found in close proximity.

117

Another version of the Llyn y Fach event was located in the Glamorgan parish of Ystradyfogwg, and centred on a pool and farm named Dyffryn Safwch. This place-name is a corruption of 'Valley of the Sow's Mouth'.[9] There The Lady-as-pig spoke for all animals, and brought them to the surface as her Other World progeny. The same archaic idea finds echoes in the medieval Welsh *Mabinogion* tales, in which the mistress of the underworld entertains Pwyll, prince of Dyfed for a year. His son Pryderi is then given her pigs, which are subsequently stolen by Gwydion the magician.[10]

Madron and Carn Fadrun

The tallest mountain on the Lleyn peninsular in north Wales is the 371 metre (1,217 feet) high Carn Fadrun. This steep-sided igneous mass is named after the Welsh mother goddess Madron, alias Modron. Her name is derived from Celtic *Matrona,* the 'Great Mother'.[11] The clouds on the mountain's summit are said to make her hat. Madron-Matrona is also related to the Romano-British *Matres* trio of mothers; and when approached from the northeast, Carn Fadrun *does* seem to have three rocky 'heads'.

Clustered on the mountain's summit plateau are the remains of numerous stone huts. Often called *Cyttiau'r Gwyddelod,* 'Irishmen's Houses', they were probably occupied at the start of Harvest ceremonial gatherings that were held on both sides of the Irish Sea.[12] (Lleyn is named after Irish settlers from the province of Leinster).

Like the seasons, Madron was always on the move. According to medieval Welsh sources she journeyed from Garth Madrun in South Wales, via Trawsfynydd in Merioneth (where the church is dedicated to a *St* Madrun), before arriving at Carn Fadrun, with her infant daughter Annun. From there she crossed the turbulent waters that swirl around Lleyn's Western tip, to reach Ynys Enlli (Bardsey), an island dominated by a hog's backed mountain.[13] Thus Madron linked mainland reality to an oceanic dream state.

Bardsey is a Welsh Other World, and a fabled *Caer* (fortress) *Sidi,* (in Irish *sid* means 'a hill of deities and fairies; enchanting, delightful', *and* 'a woman of a wondrous form'; also 'peace, goodwill, conciliation, forgiveness'). Madron shared and suited this blessed isle. The demi-god

Taliesin sang of it: 'The disease of old age afflicts none who is there... three organs play before it... And above it is a fruitful fountain; sweeter than white wine is the liquor therein.'[14]

As a *Caer Sidi* Bardsey is also a 'Turning Castle', connected to the Welsh word *sidydd,* 'zodiac', and all *its* revolving constellations, of which Madron is mistress.

Welsh *madron* means 'spinning, or giddy'. Even when stationary, Madron was the cosmic dynamo. In this sense Bardsey, like Carn Fadrun and Silbury, might be described as 'stills', advertising the Great Mother's action movie. All three sites served to project a maternal image of perpetuity, encompassing everything.

Madron had a son, the divine hero Mabon, (Welsh for 'boy'), who gave his name to the medieval *Mabinogion* tales. The earliest of these, *Culhwch and Olwen*, describes how only three days after Modron bore him, Mabon was 'imprisoned' underground at Gloucester, like the sun at night. However, thanks to the assistance of a salmon and an eagle (representing submarine and celestial realms), he was found and released.[15] In Scotland, Mabon occurs in the place name, Lochmaben, Dumfriesshire, and as a large glacial boulder, the Clochmabenstane, on the Solway estuary near Gretna.

Mabon is a Welsh version of Celtic Matrona's son, Maponos. (In Roman Britain, he was identified with the Greek god Apollo.) The effigies and inscriptions of all these deities appear together on a Roman altar at Ribchester, Lancashire.[16] At Carn Fadrun Madron re-absorbed all their fire and air into her own stony gut, prior to labouring over Mabon's next rebirth. He was the male aspect of her annual 'new fruits'.

Moel Famau, Hill of the Mothers

According to the Welsh Triads Modron also bore a daughter and another son at Llanferres in Northeast Wales. Llanferres lies at the foot of the tallest mountain in the Clwydian range. This is the 1,820-feet-high Moel Famau, or 'Hill of the Mothers', a name highly suggestive of the triple *Matres,* beloved in Romano-British times.

A folk tale places Modron at the Llanferres river ford in the role of a washerwoman, surrounded by barking dogs from the underworld.

Here the river *Alun* (Welsh 'holy') often vanishes beneath its porous limestone bed. 'I am daughter of the King of Annwfn... fated to wash here until I conceive a son by a Christian', Modron tells Urien Rheged, the legendary founder of Wales. She promises him a son if he returns in a year's time. Her triple functions, as cleanser of filth, loving mistress, *and* future mother are the three classic tasks of the Neolithic deity.[17] The Alun, like the Kennet, is involved in threshold events.

Moel Famau's summit was crowned in 1820 with an obelisk, set on a tower, both designed in the newly fashionable Egyptian style, to celebrate George III's 50-year reign. It collapsed in 1862, leaving only a stone podium behind. Even more shrunken is the previously strong belief in the *Bendith y Mamau*, or 'Mothers' Blessings', associated with the mountain. They were the now-vanished host of the *Matres'* offspring, known in English as fairies.

Moel Famau, like many other hills, could sometimes be heard *groaning* in labour. The *Annual Register* reported in 1773, 'You will scarce meet a peasant, or even a manufacturer who does not pretend to have heard the groan or sigh of a voice, rushing like a sudden wind from out of the earth'.[18] This assertion probably echoed an earlier belief in the birth pangs of the Earth-in-labour (called *Bron Trogain* in Irish) as she was delivered of the First agrarian Fruits *and* her fairy infants. As recently as 1872 a *Bye-Gones* enquirer wrote to ask: 'Are these groanings heard anywhere in Wales nowadays?' He received no reply.

The fairies were considered 'Blessings' (despite also being known as mischievous nuisances, involved in switching babies, demanding food, and trapping the unwary within their all-night dancing rings), because they acted as valued intermediaries between the current human generation, the ancestral, and the supernatural realms. With fairies (sometimes termed *Y Tylwyth Teg*, 'The Fair Folk', in Wales), divinity arrived in a timeless, low-powered mode, small enough to enter people's homes at night, for exchanges of goods or services.[19]

During six months of every year Moel Famau produced nutritious (if rough) grazing for flocks of sheep, goats and herds of cattle. Young people from the surrounding vales drove them up, at the start of May. Living in primitive huts, or *hafodtai,* they remained on the heights, milking and cheese making, before returning to the valleys in late

October.[20] This summer half tradition, which continued into the 19th Century, combined common sense with respect for the ultimate 'landlady', the pregnant Earth as provider, a figure enlarged nationwide as *Mam Cymru,* 'Mother Wales'.

Equally broad in scope, the *mam* of Moel Mamau (Welsh m's and f's mutate) means 'mother, ancestress, queen bee, a source of sustenance and protection, origin, cause, root, uterus, and womb'. Moel Famau, Modron and Mam Cymru together perpetuate some of Silbury's original meaning.

In Wales *and* England, traces of the 'Silbury' outlook survived into the 20th Century. For example, only four miles south of, and visible *from* Silbury, an August 6th to 7th assembly of shepherds took place at Tan Hill, on the escarpment of the Marlborough Downs. This lofty gathering was a market and a celebration, held on the feast of St Anne. (Drayton names it St Ann's Hill in 1620). Within that saint dwelt the pagan British goddess Anu.[21]

The Tan Hill fair site was abandoned after 1932, because lorries could not manage the climb up in wet weather. From then until 1939 (when World War II finally killed it off), the fair was moved to the field between Silbury and Swallowhead![22] But what has gone might still be understood, if not recovered, through an appreciation of Britain's evocative landforms. By their scale and latent imagery, they inspired innumerable Tan Hill-type gatherings, *and* the construction of Silbury Hill.

Mam Tor, Derbyshire

Mam Tor is the 1,696-ft-high 'mother hill' of Derbyshire. It commands a prospect over the whole Peak District, and into four counties. In prehistory, the summit was embanked and seasonally occupied from c. 1200 B.C., if not earlier, as more than 100 platforms, dug for round huts and two Bronze Age cairns indicate. Numerous potsherds and other artefacts of that and later periods have been excavated from the site.[23]

Though parts of the enclosure have disappeared due to subsequent landslips, enough survives of its outline to justify nominating Mam Tor as a topographical image, reclining in a 'birth giving' position, emphasised by the perimeter bank's outline. If this *was* the intention, as the British *Mam* name suggests, the two cairns were perhaps positioned to correspond to 'foetuses' carried in her womb.

'The vulgar theory', recorded by Camden, was that despite the Tor's many 'shivering' landslips, it never diminished in size.[24] Instead, like a living being, it (she) was able to retain her bulk. Place-names extend this superhuman metabolism to the nearby Peak Cavern (the mouth of an underground limestone watercourse), known in 1636 as Peak's Ars, or The Devil's Arse Hole.

Lady Low, Derbyshire

The entirely natural grit stone outcrop named Lady Low reclines at SK 0678, six miles southwest of Mam Tor. It was named 'the Ladie Loe' on a map of 1640, and on the ground, to this day, we can see why. The configuration looks like a 'giantess'. Her rugged head points southeast, while her eight-feet-tall 'child' (or phallic 'consort'), inclines towards the mother. The Anglian fertility goddess Freyja, wife of Freyr, was often referred to as The Lady.[25] (Anglians had settled in this area by the 7th Century A.D.) Here their 'Lady' discards her blanket of peaty soil, and sunbathes naked.

Another outcrop, half a mile southwest, is named Hob Tor, after Hob, (alias Robin Goodfellow or Puck), the hobgoblin. Supernatural male spirits apparently attended to The Lady's needs. The 'Bull Ring' Neolithic henge monument, 72 metres in diameter, which once had an internal circle of standing stones, lies only a mile north of Lady Low, and suggests that she may have drawn Stone Age worshippers.[26]

The Wrekin, Shropshire

The Wrekin is an isolated 407 metre (1,335 ft) high sow-back granite mountain. Although it lies near the edge of their large county, Shropshire people regard it as their defining 'central place', as in the Salopian toast 'to all friends round the Wrekin's circle, may our love resound'. The Wrekin provides a collective sense of identity. An expatriate, G.H. Ralphs, asks: 'Why is The Wrekin so potent a symbol in the minds of Salopians? For those born around it, or who view it with the eye of the imagination, it will always hold a certain magic.'[27]

Towards the south end of its summit crest is a granite outcrop named The Needle's Eye, comparable in form to the Lady Low feature. Here, by long-established custom, young women contemplating matrimony would crawl through the narrow gap between the huge stone 'needle' and the adjacent rocky image of the 'reclining mother'.[28] This they did from the southwest to the north-eastern or 'sunrise', side of the gap, thereby mimicking birth in both celestial and earthly terms.

Then the girls were required to drink from The Raven's Bowl, a tiny pool in a nearby rock, said never to be dry.[29] In doing so, they were swallowing the dark winter half of the Celtic year, which ended on May Day. After that the pool could become the 'Cuckoo's Bowl', a title it retained for the next six months. This division was matched by an annual May Day battle on the summit between local coal miners and farm workers, representing subterranean and surface powers. This fight was suppressed after 1826.[30]

The Wrekin Wakes, on the first Sunday in May, were immensely popular. 'The top of the hill was covered with a multitude of pleasure-seekers, with ale booths and gingerbread stands, gaming tables, swing boats, merry-go-rounds, three sticks-a-penny, and all the etceteras of an old English fair'.[31]

Professor Rhys derives the Wrekin's name from Old Welsh *gurcon,* which incorporates Irish *gur,* 'labour pains', plus *con,* 'a cone'. He adds that the prehistoric Cornovii tribe who lived in the region, and had adopted Roman habits at nearby Uriconium (now Wroxeter), located their origins in an ancestress named Gurcon.[32]

Some medieval Welsh sources enlarge the hill's name into Dinlle Ureconn, with Dinlle signifying 'Fort of Lleu'. Lleu is the Welsh form of Irish Lugh, the young god who arrives on August 1st, the start of harvest.[33] Thanks to the pairing of Gurcon and Lleu, The Wrekin (which remained in Welsh hands until the 8th Century) has etymologically convincing 'Mother goddess and Son' credentials. Moreover, as we have seen, Gurcon displays on her hog's back profile the monumental image of herself as a superhuman goddess-in-labour, carved by Nature out of granite, with a passage-way offering the means for young women to identify with her annual travail.

Pontesford Hill and the Golden Arrow

Seventeen miles west of the Wrekin, the 520-metre-high Pontesford Hill rises 'like a crouching lion', writes Burne;[34] (or, in a county without lions, perhaps a pregnant giantess?) Lleu may also have left a trace of his former presence on Pontesford Hill in the traditional 'Golden Arrow Hunt', even after this rite had been switched to a Christian Palm Sunday date.

As Burne reported: 'People believed that the golden arrow had been dropped in days gone by, and that the search for it was ordained by a good fairy, as the condition on which she would undo some unknown injury, curse, or spell, inflicted by a demon. But to be successful the quest had to be undertaken at midnight by a young maiden under 20, who was the seventh daughter of a seventh son. That would ensure good fortune for a year'.

Was this elusive 'arrow' originally intended as the first sunbeam of the year's May Day summer half, and a precursor of the first ripe corn stalk, which was Lleu's staff and emblem? A seasonal reference seems probable, since after villagers had gathered on the hill for dancing and drinking, there was a race to pluck the first spray from an ancient and reputedly haunted yew tree growing nearby, perhaps as a token of the dark winter half that was about to end.

The second race was down the very steep east end of the hill to the Lyde Hole, where a brook falls into a crater, believed to be bottomless. Whoever could run at full speed into the Hole from the hilltop and dip the fourth finger of the right hand into the water was *bound* to marry the first person seen thereafter.[35] This resembles the Wrekin's Needle and Raven's Bowl ceremony, but with a dangerous underworld dimension.

Prior to the spread of Protestant distaste for Britain's erotic landscapes, Lyde Hole may have served as Pontesford Hill's 'vulva'. From an Indo-European perspective, the golden arrow finds an equivalent in the god Siva's fiery arrow of ultimate energy, which can only be stabilised within the vulva of his voluptuous wife Parvati, the Lady of the Mountain, and daughter of the world's central mountain.[36] Without that union, Siva's arrow would have turned the world to ash (a fate now rapidly approaching).

Is this Siva-Parvati comparison with Pontesford far-fetched? Pontesford and India are much closer than is Paviland from Siberia, whose Palaeolithic figurines Aldouse-Greene so convincingly links together.[37] Ever since then, Eurasia has shared a pool of myth, deeper than Lyde Hole.

The Banbury Stone, Worcestershire

In a hollow on the 997-ft top of Bredon Hill, an outlier of the Cotswolds, sits an enormous honey-coloured limestone boulder, named the Banbury Stone. It is roughly 10 ft high and 60 ft long, yet people believed that at midnight the Stone travelled a mile downhill to drink from the river Avon, and then returned.[38]

As with Derbyshire's 'Lady' and The Wrekin's 'Needle's Eye', a two-ft-wide gap, through which initiates could pass, separates the Banbury Stone's 'groin' from her eastern 'child', so offering another version of the goddess in labour.

A landslip, that occurred about 1800, close to the Stone's foot, exposed a large quantity of stored wheat, 'perfectly formed, but which crumbled to dust if touched'.[39]

Another aspect of the Banbury Stone's 'underworld' was described in Dr Derham's *Physico-Theology* of 1712. He wrote: 'Near the [Bredon] precipice facing Pershore, in or near the old fortress, is a cave lined with stalactitical stones... On the top they hung like icicles, great and small, and many lay on the ground... as an exudation of some petrifying juices out of the rocky earth there.'[40]

Whitsuntide games were held at the Banbury Stone.[41] They included wrestling, shin kicking, quarterstaff contests and bare-knuckle boxing. The Banbury Stone itself was danced around; hence its name may derive from Latin *ambire,* 'to go around'. Others connect the boulder's colour with 'amber', a stone once regarded as helpful in childbirth; and 'Banbury' was locally pronounced 'Bambury'.[42]

The Stone continued to be viewed seriously in Christian times. On Good Friday many pilgrims climbed the hill to kiss it.[43] By this act, the pagan 'Lady' merged with the Blessed Virgin Mary. Here, as at Silbury, water was an essential element in the Banbury Stone's fecundity. Veneration of a spring, 200 feet below the Stone, continued into Christian times. A chapel to St Catherine was built near the outflow.

Roseberry Topping, North Yorkshire

The Danish word *toppen,* 'pointed,' explains the place-name's second element, attached to this 1,057-ft-high mountain. It rises in isolated splendour on the edge of the Cleveland Hills near Middlesbrough. A Nordic connection also underlies the name Roseberry, a corruption of 'Othenesberg', from Odin, the chief Norse god.[44] His *Aesir* pantheon was hostile towards the fertility goddess Freyja, yet he was sometimes represented as her husband!

They both had links with the underworld that have seeped into this hill's folklore. For example, when a sick person's shirt was placed in the well at the mountain's foot, 'if yt floated aloft yt anounced recovery to the partie, but... if yt sanke, there remained no hope of health'.[45]

A local legend tells of a Danish princess who dreamed that her infant son would drown. In order to thwart the presentiment she climbed with him to the summit of Roseberry, high above any dangerous river or lake. However, when she fell asleep on that warm summer's day, he drowned there, in the tiny mountaintop spring.[46]

Perhaps this tale contains a faint memory of harvest sacrifice, acknowledging the two-way traffic between *sid* and surface worlds, which is also apparent in the Roseberry Topping folk custom of 'threading the needle'. As at The Wrekin, so on this mountain there was 'a clefte in the rocke'. Here it was called St Winifryd's Needle, 'whither blynde devotyion led many a syllie soule, not without hazard of a breaknecke tumblinge, while they attempted to put themselves through a needlesse payne, creepyinge through that Needle's eye.' (St Winifred is the English form of Welsh Gwenfrewi, whose decapitated head rolled downhill to create the still-revered well at Holywell, Flintshire. Divine powers from east *and* west converged on Odin's hill.)[47]

Roseberry's many 'sillie soules' had inherited a prehistoric need to identify with the sanctified Mother Earth, for they were the offspring and future replicators of that mountain's fecundity. Accordingly, they undertook to be physically 'born' from the rock.

In a similar, but reversed vein, a hoard of prehistoric Bronze Age implements, including socketed axes, gouges and hammers, and a bronze sheet with crescent moon-shaped openings, was found hidden beneath a boulder, close to the summit. The earth's precious ores were returned with interest.[48]

Cruckbarrow Hill, Whittington

On the western outskirts of Worcester, at Junction 7 of the M5 motorway, this 28-metre-high tump is a major local landmark, which can be seen over a wide area. With a name derived from British *crug*, 'hill', it has been compared to Silbury Hill by Jabez Allies. He attributed Cruckbarrow's 'extreme regularity of contour to artificial means'.[49] The Geological Survey classifies Cruckbarrow as 'artificial ground', an assertion that has yet to be tested.

Cruckbarrow was an important gathering spot. F.T. Spackman recalled that until 1914 'there has always been, once a year, a great concourse of Worcester people on Crookbarrow Hill, with swings, roundabouts and other accessories for merrymaking, to the very great inconvenience of the owners of the estate. This annual exodus of citizens [from Worcester] occurs on Good Friday – perhaps the most sacred of all the fasts observed by the Church. [Yet] so firmly established is the custom, so great are the throngs of people who attend, that… no opposition, as far as I know, has ever been offered to the merrymaking'.[50] This gathering died out in the 1930s.

A Neolithic flint scraper and Roman coins found at the site suggest that in prehistoric times and beyond, Cruckbarrow might have attracted 'harvest hill' rites. As might be expected of a possible First Fruits location, Cruckbarrow, alias 'Cooksey Hill' was involved in the task of converting the white stones that lined The Tything (a Worcester street), back into the men and horses that they had once been. A folk tale asserts that for this transformation to occur, a new loaf of bread had to be placed at midnight on each stone, precisely *'when a half moon was rising over Cooksey Hill'*. But the only man brave enough to try was terrified when one stone that he had loaf-crowned changed into a gigantic, rearing horse. The man ran off, failing to complete the task.[51] Does this crumb of folklore fall from a Cruckbarrow 'First Loaf' rite?

Five Four Stones Hill, Clent

This prominent isolated hill, reaching 315 metres (1,033 ft), lies at the northwest end of the Clent hills in Worcestershire. As a natural counterpart to Silbury Hill, it attracted local 'start of harvest' rites. Like Silbury, Clent was named as the centre of a 'Hundred' – a group of Anglo-Saxon parishes, where law courts were held.[52] (Both Demeter and Ceres were regarded as originators and disseminators of law and order.)

The hamlet at Clent's eastern foot was the scene of a Fair, held on three consecutive days in late July, and endorsed by royal approval in 1254. Horses, cattle and sheep were traded, as were linen cloth, cheeses and pedlars' wares, with the third day given over to pleasure. Records show that this well-attended market and celebration persisted till at least 1784 'and at nearly every house were sold roast beef and ale… with accommodation for man and horse.'[53]

The Fair was linked to a 'Crab Wake', which took place on the first Sunday in August, with the two events spanning the August 1st quarter day. Taken together, they may represent an instance of 'Sillie Season' continuity. (In linguistic terms, 'sillie' derives from Middle English *selli*, 'a wonder, marvellous, supernatural', and from M.E. *seli*, 'blessed, prosperous, happy'.) The Crab Wake took its name from 'a strange custom of pelting each other with crab apples'. Two barrow loads of these were deposited in the churchyard and flung at the parson and congregation when they emerged from church.[54]

By the mid 19th Century the Fair had shrunk to 'a few stalls erected by the side of the churchyard, in the narrow road by the brook, at which some trifling articles were sold, the chief commodity being cherries'; so Amphlett recalled in 1890. As for the Wake, 'being accompanied by great disorder on the hills, it was soon suppressed, and the Fair vanished with it'.[55] The rough behaviour may have derived from the ritual 'faction fights', like those held on The Wrekin, that were also a feature of the otherwise notably peaceful Irish Lughnasa gatherings.

At Clent, history and mythic violence combined with the murder of an Anglian boy-prince, Kenhelm of Mercia. Here, in A.D. 819, he was supposedly killed and buried by his half sister Quendrida and her paramour. 'O Kenhelm, martyr, glory of Mercia, exceedingly sweet is thy memory, for thou surpassest the honeycomb by the delight of thy sweetness', runs a medieval Latin hymn in his praise.[56] Since Anglian royalty believed themselves descended from their own pagan gods, the boy Kenhelm fell from a divine family tree.

As befits a supernatural harvest child, cut down like John Barleycorn in order to initiate the ripe crop's reaping, Kenelm's name derives from O.E. *cennan*, 'to beget, to conceive, to bring forth'. He is the symbolic child of O.E. *cenning-tid,* 'the time of bringing forth', subsequently Christianised into Lammas, 'loaf mass'.

After his clandestine burial, 'a white cow... belonging to a certain widow, hastened down from the mountain-top [Clent's summit], to the grave, and there it remained, nor could any force drive it away', says Richard of Cirencester. In Nordic and Germanic myth the white cow is Audumla. She emerges from ice at the beginning of time, and licks into life the first giant Ymir, progenitor of humankind. Then her udders yield prodigious streams of milk that become the world's rivers.[57]

In Celtic belief, a comparable white cow comes from under the *sidh* as a sacred avatar of the goddess. Pastoralists sang of, 'a white cow on the mountain, A fair white cow; She goes east and she goes west. And my senses have gone for love of her. She goes with the Sun, And he forgets to burn, And the moon turns her face with love to her'.[58]

At Clent, when Kenhelm's body was eventually exhumed, a dove flew from the pit, carrying an account of his murder to the Pope in Rome. Then a spring burst forth where the white cow had lain down on his buried body. Directly over the spring a church (rebuilt in the 12th Century) was constructed. As Richard reported, 'the stream still flows and has dispensed healing power on many who have tasted it'.[59] In 2002, devoutly hung ribbons still dangled from branches of an ash tree, overhanging a subsidiary well, 20 yards downstream, the original source having been diverted in the 19th Century, because it made the building damp.

Notable hill, holy spring, sacred cow, sacrificed harvest boy, August quarter day festival and market; here the elements of a Neolithic First Fruits event retained a matrix into which an alternative Christian theology, other legends, historical particulars and social changes were absorbed. In England nothing is absolutely discarded, yet everything is modified, to a degree where the different layers and connections are often hard to decipher. Yet compared to the 200 million-year age of the hill itself, *all* these uses, including a likely Neolithic interest, combine to make a single, recent conglomerate, which can be enjoyed in all its confusing variety.

For the thousands of Birmingham and Black Country people, young and old, who climbed Clent Hill on the night of June 6th, 1977, to celebrate Queen Elizabeth's jubilee, the communal sense of attachment to a special place for a shared reason was palpable. Around the summit bonfire, a people and their queen were incorporated into a national chain of beacons, blazing from Land's End to John o'Groats, with the neighbouring fires, on Malvern and Cannock Chase, plainly visible. Thus 'ordinariness' there and then became *extra*ordinary. Hill, fire, monarch and her hilltop throng, together generated an unexpected encounter with the sublime.

In that orange glow, among darting silhouettes, something of the spirit of Silbury almost (and briefly) came to life; and it does again (for some readers) in the work of the poet James Thomson (1700–1748). He was a frequent visitor to Hagley Park, which laps against the north flank of Clent Hill. There he composed much of his best poem, *The Seasons.*

The Seasons is a reminder that even in an Age of Reason, such as his (and ours), things can still be seen in the round, despite our concentration on 'linear progress'. Of this poem Dr Johnson wrote: 'He brings before us the whole magnificence of Nature, whether pleasing or dreadful… the gaiety of spring, the splendour of summer, the tranquillity of autumn, the horror of winter. Our thoughts expand with his imagery, and kindle with his sentiments'.[60]

Equally radical, in their challenge to the monopoly of *linear* time, are the architectural follies, commissioned by Lord Lyttelton of Hagley Park during the mid 18th Century. These include four 'druidical' standing stones, erected on the summit of Clent Hill. On the lower slopes, arcades transported from the ruins of Halesowen Abbey are attached to a mock medieval 'castle'.

Nearby, a Greek Doric temple – the first ever seen in Britain – was built in 1758. Its front was a miniature copy of the 4th Century B.C. Theseion temple in Athens, dedicated to Athena and to Hephaestus, the god of divine fire and metalworking.[61] Sited on the fringe of the industrial Black Country, but with a view of rural England, Lyttleton's 'temple' fits, not least because Clent and the neighbouring hamlets were noted for making steel sickles and scythes, the tools of agrarian continuity and peace.

In their beautiful setting, the Hagley follies, gathered beneath Clent Hill, implicitly speak in favour of an all-time companionship, defying mortality and mediating between differences. If that ambition is folly, then so is Silbury's ability to address the needs of every era, without chronological impediment. Thanks to English Heritage she is still intact, ready to greet us again.

Harvest Hills and Saints

Faced by medieval Christian disapproval and the threat of prohibition, many pagan hilltop ceremonies in England survived by adopting a saintly cover story, such as Saint Catherine's. Her hagiography states that she was a princess of Alexandria, martyred on a wheel, before her body was carried by angels to the top of Mount Sinai and buried there. Her legend was brought to England by returning Crusaders.[62]

From the 12th Century onwards, numerous English hilltop chapels were built in her honour, as at Barmby Moor, Yorkshire; Houghton on the Hill, Leicestershire; Guilford, Surrey; Chale, Isle of Wight; and Milton Abbas, Dorset.

At Abbotsbury, Dorset, her chapel tops a hill scored with seven Bronze Age cultivation terraces. Into the south doorpost generations of young women have pressed their elbows and knees, and worn deep hollows there, while appealing to the saint on St Catherine's Eve, November 24th, for a husband:

'Sweet St Catherine, send me a husband, a good one, I pray;
But arn a one better than narn a one
O St Catherine, lend me thine aid,
And grant that I may never die an old maid'.[63]

131

This version is from Cat and Chapel Hill, Piddletrenthide, Dorset. The timing of the chant allowed nine months for the desired mating to turn pregnancy into a Lammas birth. St Catherine provided a hill for all seasons, as her rotating wheel implies.

But Christianity was not always so accommodating. Many former Harvest Hills were assailed and nullified by the militant archangel, St Michael, Satan's adversary. For example, a late Saxon church and a surviving 13th Century tower, both dedicated to him, were built on top of Glastonbury Tor, Somerset. The Tor is a steep-sided, isolated hill, more than 500 ft high, overlooking the marshy Somerset Levels, and visible for 20 miles around.

Worked flints from every phase of the Stone Age have been found during excavation of its summit, though this does not necessarily support the view that the dramatic terracing around the Tor's flanks are a maze, created in the 3rd millennium B.C. St Collen is reputed to have made a cell on the Tor in the 7th Century and had a vision of Gwynn ap Nudd, King of The Underworld, sitting in his abysmal castle, where a great feast was in progress.[64] (Was it as a concession to the Tor's pre-Christian associations that an image of a woman milking a cow was carved in low relief onto the church tower?)

St Michael had much more work to do in Somerset. His churches crown Burrow Mump's conical hillock and St Michael's Hill, Montacute. In Devon, another of his chapels, built by the monks of Tavistock, tops the precipitous Brent Tor. He also claims several Cornish conical hills, including the isle of St Michael's Mount. Throughout England, the frequency with which St Michael appears on likely 'harvest hills' shows how hard it was to eradicate worship of the prehistoric harvest deity.

Even *within* Christianity, harvest spirits could not be denied. For example, over the north door of Stoke-sub-Hamdon's church (only a mile from Montacute) a 12th Century tympanum carries the parishioners' corn-growing concerns sky-high. It shows Sagittarius firing his midwinter arrow at Leo's August rump, with their names carved on the firmament's edge, to make doubly sure he does not miss. Between them, a tree grows from a miniature 'Silbury' creation mound, thereby drawing underworld and celestial vault together. The Neolithic occupants of the Hamdon Hill enclosure, which overlooks the church, would have enjoyed this Norman carving.[65]

Clegyr Boia, Pembrokeshire

One mile west of St David's, Pembrokeshire, a rocky outcrop, approximately 45 ft (13.7 metres) high, rises above fertile fields. Named Clegyr Boia, its 300 yard (274m) long outline, seen from the southeast, makes the classic profile of a reclining 'Lady'. On the undulating summit, post-holes for Neolithic houses have been found. Both rectangular and oval, these timber and daub dwellings date from the mid 4th millennium B.C.[66] The inhabitants were believed to have migrated from Lough Gur (Irish *gur,* 'labour pains'), a well-known sacred site in County Limerick.[67] At Clegyr Boia the villagers tilled the surrounding fields, kept cattle, ate, drank, prayed, sang, danced, made love, gave birth, slept and died on their deity's stony breast. They were the human representatives of the Irish *Tuatha de Danann,* 'People of the goddess Anu' (Danu).[68]

Occupation of Clegyr Boia continued into the 6th Century A.D., by which time St David had arrived, intending to set up a monastery nearby. His plan, legend records, annoyed the wife of Clegyr Boia's chief. To drive the saint off, she sacrificed her own daughter, Dunawd. When that had no effect, except to cause a spring named Dunawd to erupt, she encouraged her servant girls to utter obscenities and to dance naked in the river Alun before the monks, who were 'disturbed, but did not retreat'.[69] The Alun valley is still known as Merry Vale, and the land around Clegyr Boia continues to be fertile; it grows cabbages. The monks have gone, but St David's cathedral flourishes.

Carngyfrwy's Bluestones

Twenty-three miles east of Cregyr Boia, on top of the Preseli Mountains, there is a volcanic dolerite outcrop named Carngyfrwy. From the scree around this spectacular natural 'effigy', evocative of the reclining goddess-in-labour, the famous 'bluestones' were collected and somehow transported to Salisbury Plain, where they form a U-shaped setting within the sarsen ring of Neolithic Stonehenge.[70]

The bluestones are Carngyfrwy's natural scree 'offspring'. White egg-shaped spots are a notable feature of this rock. At Stonehenge, 17 miles from Silbury, the U-shaped layout of the bluestones makes a well-known symbol for the womb, as seen on Minoan figurines and in much British folklore concerning women's aprons. (In Wales, many a prehistoric cairn is explained and named as a load of boulders dropped from a giantess's apron.)[71]

The Stonehenge bluestones also emphasise the island-wide nature of the 'goddess and her progeny' theme. During the Neolithic, and indeed throughout the succeeding Bronze, Iron, and Roman periods, Wiltshire and West Wales apparently spoke the same earth-based visual language. In A.D. 540 the Welshman Gildas, a founder of monasticism here, attacked this long-established earth worshipping tradition, writing: 'I shall not speak of the ancient errors ...that bound the whole of humanity before the coming of Christ... I shall not name the mountains and hills and rivers, once so pernicious, [but] now useful to human needs, on which... a blind people heaped divine honours.'[72]

His conviction that the new monotheism had brought an end to hill worship was premature. In 1886 Professor Rhys noted that 'the echoes of a feast or fair on August 1st have not yet died out in Wales. In Cardiganshire till recently the shepherds had a sort of picnic on the hills. A farmer's wife would lend a big kettle for a good broth and everybody present had to put some fuel on the fire'.[73] With those simple gestures several thousand years of August hill worship flickered and finally died.

Silbury and the Birth Chairs

Its rites now over, for some observers the Silbury image is hard to recognise, because although the hill and moat together combine to present a clear 'squatting in parturition' pose, most of the figure is laid out on a *horizontal* plane, rather than sitting erect. This was necessary because the effigy was largely drawn in water. However, many other

Neolithic and later 'squatting goddesses' in Britain *do* employ a vertical format, though on a much smaller scale. A few surviving examples are given here:

A native effigy, carved in sandstone at Roman Caerwent, Gwent, bears (in both senses) a sapling fir tree as a symbol of fertility. She *is* the ground into which the tree is rooted, *and* the rocky throne upon which she sits.[74]

In other instances, the chair, stool or throne could, by itself, represent the divine birth event. The six-feet-tall Neolithic menhir at Stackpole Warren, Pembrokeshire, is one of many in Britain with a ledge that suggests an 'abstract' rudimentary image of a seated supernatural figure, and one that invites a human sitter. Similarly, the natural seat within the south portal stone at the Avebury henge, called the Devil's Chair, was until the mid-20th Century, where the May Queen sat on May Day morning.[75] In this way the deity encouraged humanity to join in, and engage directly with the world-creating process.

The supernatural dimension is also found in the medieval Welsh description of a seated maiden in *The Dream of Macsen Wledig*, when the emperor sees a maiden, whose beauty is brighter than the sun, sitting 'in a chair of red gold'. She is dressed in 'vests of white silk, with clasps of red gold at the breast; and a surcoat of gold brocaded silk upon her ...and a frontlet of red gold on her head, with rubies and gems on the frontlet, and pearls alternately, and imperial stones; and a girdle of red gold around her'.[76] Macsen's intoxication is partly distilled from the Celtic sun goddess tradition.

The deity-*as*-chair was regarded as the source of eloquence. Therefore the 'chair' was awarded to the person who could best access its inspiration; hence the bardic chair, awarded annually to the finest poet at the Welsh national *eisteddfod* (a word meaning 'sitting together').

Likewise the Stone of Scone, for seven centuries embedded under the coronation chair at Westminster Abbey, provides the physical link between Britannia, the superhuman rocky island, and her enthroned human representative, the new monarch, who interpreted the divine word into law. All these manifestations may be seen as derivatives of a Neolithic squatting deity prototype, of which Silbury is the grandest surviving example.

The Gop

Surpassed only by Silbury Hill, The Gop is the second largest artificial prehistoric mound in Britain, and believed to be Neolithic.[77] This 46-ft-high monument, built of rough limestone blocks, is a steep-sided, flat-topped oval in shape. Its long axis measures 223 ft, aligned northwest to southeast. Somewhat damaged by stone robbing, Gop stands on a 774-ft-tall hilltop in Gwaenysgor parish, Denbighshire, at SJ 085724, overlooking the mouth of the Dee estuary.

The name Dee (Latin *dea*, 'goddess') is a generic form of the river's original Welsh title, Dyfrdonwy, 'Waters of the Goddess Don'. Don was the mother of an entire Welsh pantheon of deities, listed in the *Mabinogion*, including Lleu and the farmer god, Amaethon.[78] Gop may be regarded as their collective birth mound and supernatural home, equivalent to an Irish *sid*. *Y Gop* may be a shortened form of *Cop-y-Goleuni*, 'Shining Crown' – a name in use until c. 1860.[79]

When Boyd Dawkins dug vertically and horizontally into Gop in 1861 he encountered no burials. Like Silbury, it was not a tomb. However, only 40 yards (36.5m) to the south, he discovered, within a limestone cave, a Late Neolithic polished axe and skeletal remains of at least 14 people, surrounded by deliberately placed boulders. There were also some woolly rhino, bison and reindeer bones of Palaeolithic age, and some Bronze Age artefacts.[80]

Boyd Dawkins was struck by the close resemblance of the cave's human skeletons to the 'small, dark Welsh who are still to be seen in the fairs and markets of Denbighshire' and concluded that they were descended from the original Neolithic stock that had been settled in the area through more than four millennia.[81]

Willy Howe: A Ruined 'Silbury'?

Willy Howe, at Thwing, is the largest 'barrow' in Yorkshire's East Riding. Its shape, now ruined by treasure hunters, was a flat-topped, grass-covered, chalk cone, 25 ft (7.6 metres) tall and 125 ft (38 metres) in diameter at base. Opened by Lord Lonsborough in 1857, 'there was no appearance of a body ever having been interred, yet animal bones found therein proved that any human bones could not have completely decayed away'. Instead, beneath the cone's centre was an oval, or vulva-shaped shaft, sunk more than 13 ft into bedrock, and filled with layers of earth and chalk.[82]

The archaeologist William Greenwell noted that 'on more than one previous occasion ...I have found mounds lacking any sign of interment'.[83] He cites one at Folkton, 60ft (18 metres) diameter, where 'a flat mound of earth was first made, and on this a smaller mound, 42 ft diameter, of chalk and a little soil was placed. This was enclosed by a 1 ft-high wall of chalk slabs, resting on the primary mound'. As at Willy Howe, beneath the entire structure, a vulva-shaped pit (in this case eight feet by five feet, and four feet deep) had been sunk, with 'not the slightest indication that a body had ever been buried in it; very difficult to explain'.[84]

Midlothian Harvest Hills

In the villages of 18th Century Midlothian, Scotland, 'the celebration of the Lammas Festival was most remarkable', wrote James Anderson in 1792. Each community built a 'Lammas tower' at some conspicuous place, close to a spring or well. Each tower was built during July and was usually approximately 1.22 metres (four feet) diameter at base, 2.4 metres (eight feet) tall, tapering to a point and topped by a 'flag', which was a large table napkin, decorated with ribbons. The structure was then closely guarded to ward off attacks from neighbouring villages, prior to the 1st of August.

Then, dressed in their best clothes, people marched to the tower, blowing horns and 'making merry', before a meal of bread and cheese was eaten at the tower's foot, and water, fetched from the nearest well, was ceremonially drunk.[85]

Similar, but *prehistoric* turf stacks, found at West Heath, Sussex; at Six Wells, Llantwit Major, South Wales; and at Crig-a-mennis, Cornwall, indicate that the roots of the Midlothian tradition probably developed with the beginning of Neolithic agriculture in Britain.[86]

As this chapter has attempted to show, Silbury does not stand in isolation. Rather it belongs among a great company of long-venerated features. They include (i) entirely natural topographical 'pregnant goddess' effigies, regarded as sacred, (ii) hillocks and rock stacks modified by humanity to emphasise their Harvest Hill character, and (iii) entirely artificial structures, created to simulate and wait on the Earth deity in labour. All three types served as assembly places, where the Earth Mother and the deities of Harvest-time were honoured, and their stories ritually re-enacted in order to initiate successful reaping and gathering of the ripening crops.

Cesair, Harvest Hill and Lake

MacNeill's account of Lughnasa, the start of harvest festival in Ireland, lists no fewer than 95 hill and mountaintop sites where August 1st celebrations were known in 1942. In addition, near Boyle, Co. Roscommon, there is a 20-ft-high flat-topped mound reputed to be the grave of Cesair, Ireland's *magna mater*. According to the medieval *Lebor Gabála*, she was the daughter of Bith (whose name means 'cosmos'). After causing a flood (her breaking waters?), she creates Ireland, its hills and people (Irish *cessair,* 'to carry, to bring forth').[87]

The Irish August 1st rites may be seen as a vestigial repetition of Cesair's primal nativity. As at Silbury, in several locations these celebrations involved both a hill *and* a lake. For example, on the shores of Loughsalt, Co. Donegal, there was a gathering on the first Sunday before, and the first after, August 1st, until c. 1910, as a man from the district recalled. 'They were very important festivals… eagerly looked forward to.' Young people from four parishes, including fiddlers and melodeon players 'would gather in the evening for dancing, singing, with courting couples very much in evidence and marriages arranged'. Some people came to enquire after lost or stray sheep. Then the steep white quartzite mountain, Cnoc a' Liathain, that overlooks the lough, was climbed. Ripe bilberries growing on the summit were threaded onto long stalks of grass, and carried home to give to older relatives.[88]

Lowbury Hill, Oxfordshire

On the summit of this 185-metre-high (609 ft) chalk hill that gives magnificent views over fertile downland, a sequence of potsherds has been found, running from beakers, c. 2000 B.C., through to Bronze Age, Iron Age, Roman and early Saxon vessels. In addition, before A.D. 100 a Romano-Celtic walled rectangular sacred enclosure, 48 by 60 yards

(44 x 55 metres), was constructed on the hilltop. Votive offerings left there include 57 brooches, 34 pins, four tweezers, 22 finger rings, seven needles, eight bracelets and 1,019 Roman coins.[89]

To the archaeologist M.G. Fulford the many sherds of *mortaria* (one stamped with female images) and flat dishes discovered at the site imply the ritual preparation and consumption of food, as part of the pattern of worship. Further, the ceremonial 'slots' or deep 'first furrows' that Fulford found within the divine enclosure, and the oval shaped pits filled with dark loam, may be further signs of Earth deity rites there.[90]

That these were orientated towards the Harvest Mother is confirmed by the recent find nearby of a bronze sceptre head, incorporating three images of the Great Goddess, displaying her maiden, mother, and old woman forms. These spring from leafy volutes. (Unfortunately the plump 'mother' has lost her head.) Only 27 miles east of Silbury, this three-in-one cluster of deities 'were deeply involved in the whole of Nature', writes the Roman Art expert Martin Henig. He envisages the sceptre being carried in religious processions, and relates the object to three similar 'tipstaffs' found within the Lowbury summit enclosure in 1914.[91]

As if to emphasise a Lowbury – Silbury resemblance, a bronze bust of Diana, the Roman goddess associated with the moon, mountains, and women-in-childbirth, has been unearthed less than half a mile from Lowbury.

The 7th Century pagan Saxon barrow, constructed a few yards from this Romano-British temple, is interpreted by Fulford as a Germanic newcomer's wish to achieve continuity with the spirit and native purpose of the site. This may also explain the burial of a middle aged Saxon woman in the bank of the temple compound.

'WONDER' OF THE PEAK

The deepest entirely natural pothole found in the limestone rocks of the south Pennines is Eldon Hole, near Peak Forest. The 17th Century philosopher, Thomas Hobbes, said, 'its mouth is obscene' being of 'a cunnoid form, shaped like a woman's privities'; and he pretended to keep this a secret: 'Tell me, tell't in my ear. Whisper't that none but thou and I may hear.'[1] The generalised belief in the land as a living female entity is here explicitly and dangerously apparent, on both a natural and perhaps a *super*natural level.

The name Eldon derives from Anglo/Saxon Elvedon, retained till the 13th Century and meaning the haunt of elves,[2] the benign attendants of the Earth Goddess, here termed Don, equivalent of Irish Anu, head of a pre-Christian pantheon.[3]

Given these links, the hole was therefore said to be unfathomably deep, because it led to, and was part of, the immeasurable Other World, long believed to underlie and sustain 'ordinary, surface reality'. From its 442 metre (1,450 ft) top, Eldon offers a gaping mouth between different *kinds* of worlds, and so is likely to swallow 'common sense'. Eldon is therefore listed as a Wonder of Derbyshire's Peak district.[4]

In the 16th Century, after a poor peasant for a pretty price, was lowered on a rope, 'one hundred yards-long' into this abyss, he returned to the surface raving mad, 'with staring eyes, and incapable of "normal" speech, vented something none could understand', and then died within eight days, according to the Earl of Dudley.

Similarly 'In the Middle Ages madness resulted from a dramatic debate, in which Man confronted the secret powers of the world',[5] says Marcuse. He ascribes such mental disturbance to 'the recovery of the unconscious buried within history; a re-turn from alienation, in which, instead of following the ascending curve to the spirit of transcendentalism, the individual completes a circle through the neglected Earth, to rediscover a Holy Ghost within the body of the rock.'[6]

Local folklore here records that when hunting for her goose a poor old woman accidentally drove it into the pothole, to her great sorrow, from where it emerged days later, at Peak Cavern, Castleton, more than two miles away. This cave is often called The Devil's Arse Hole. During its underground journey to that cave, the goose's feathers had been singed off, for, in Christian terms, it had passed through Hell.[7] The connection of Eldon's presumed underground waterway to the Devil's Arse[8] also serves to place The Feminine, as a category, in Hell.

Ever since the appearance of Judaic Eve, women have dwelt in the realm of Sin, and been integrated with the embodiment of negative power – The Devil.[9] Consequently the old woman's goose was redefined as the Devil's turd, though geese had previously been sacred to the pre-Christian British.[10]

The Derbyshire poet, Charles Cotton, emphasised that the Devil's Hellish quality extended to Eldon Hole. After standing at its lip, he wrote: 'I must declare, My heart still beats, and Eyes with horror flare', as at 'The brink of Hell', adding that: 'A stone thrown in, A kind of sighing makes', and 'seems to hiss, Like the old serpent in the dark abyss'.[11]

Likewise the old deities, often female, were, by definition Hellish. When Cotton drew up his rope up from the abyss, he found it tied into 'scores of curious knots, made wondrous fine',[12] as if elf spinners were also still at work below. This was despite the superimposed Christian sediment, including Dr Leigh's belief that the pothole and its river dated from the time of the Biblical Flood[13], and W.F. Ainsworth's assertion that the cave's arches were made in an ecclesiastical Saxon style.[14]

Eldon Hole stands for the layered complexity of European culture, and as Kathleen Raine says, 'It is difficult for us to realise that our unspoken assumptions about the world, the very foundations upon which we build it, differ from those of others, for this assumed ground is the very thing that we cannot discuss, since it does not register on our instruments.'[15] Yet it, and the dilemma, remain all encompassing.

In the early 18th Century, the traveller Celia Fiennes reported that: 'Severall attempts have been made to fence the [Eldon] hole round, with a stone wall, but what they build up in the day, would be pulled down in the night, and so it is vaine to try securing it round – so the people tell us'.[16]

The failure to shut off the hole suggests that continuing contact with the 'foreignness' held therein, made an indispensable contribution to people's surface 'truths'. And/or at least that Other World forces still craved interaction with the Earth's surface 'reality', and could not be denied.

However, to Newtonian advocates of a purely mechanical cosmos, the last traces of this alternative vision had to be eliminated *by measurement.* Consequently, in 1770, J. Lloyd came from the Royal Society in London and was lowered into the hole by eight men, using two ropes of nearly 40 fathoms (73 metres) in length.[17] (He had often read accounts of Eldon's bottomless nature, and was aware that 'a farmer had spent some days throwing many loads of stone into it, to no effect'. In 1636, Hobbes had reported that the hole 'would have swallowed vast mountains tall'.)[18]

At the surface Lloyd measured the 'woman's privities' as 30 yards (27 metres) long by 8 yards (7 metres) at their widest. At 80 metres underground, he found himself on a ledge of recently tipped-in stones, below which a narrow entry led to a stalactite encrusted domed cavern; while from its floor another vertical shaft led downwards, for a further 80 metres, to an underground waterway, presumed to lead to the Devil's Arse.[19]

Fifty years earlier, Cotton, 'By half the Peake surrounded', had measured the vertical drop as 'eight hundred and four score and four yards', thereby showing that old poetry, despite local witnesses, must give way to New Science.[20]

Yet less than a century earlier, Samuel Purchas had outlined an entirely different educated norm, in his book *Microcosmos, or the History of Man*. He asked: 'Is not the Haire as Grass, The Flesh as Earth, The Bones as Mineralls, the veines as Rivers? The two Eyes akin to the greater lights (of Sun and Moon)'? Note his repeated use of 'as', as if an earlier complete unity was, by 1619, already slipping away into mere simile or resemblance. [21] Yet his contemporary, Sir Walter Raleigh, could recall a time when the ground gave rise to giants,[22] and illustrations to Drayton's topographical poem, *Poly-Olbion*, of c. 1620, populated every stream in England with a female deity, and placed crouching female figures under both Eldon Hole, and The Devil's Arse.[23]

However by 1790, as William Blake recognised, the Battle over Eldon Hole seemed to have been won by the New Science. Consequently 'The Starry Heavens are fled from the mighty limbs of Albion and the phenomenal world, emptied of spiritual life, had been converted into a "soul-shuddering vacuum", with humanity alienated from a dead Nature'.[24]

Top: A man being lowered into The Hole. (after Richard Muir, c.1810).
Bottom: The Eldon Hole Great Chamber and underground river.

From 1956 onwards, the very hill into which Eldon Hole descends was half consumed by a massive road stone quarry, which has now been denied permission to expand further.[25] The hole has been saved for the nation, but where does it lead?

The name Peak Cavern name was invented to please Queen Victoria, who, in 1880, was about to attend a concert in the Devil's Arse Hole, when the organisers decided on the alteration.

Yet to this day the original title is still in use, partly because it accurately describes the flatulent sounds that air makes, when pushed through the caves by an underground stream in spate. This extensive cave system is approached from Castleton village by a 280 ft (85 metre) limestone gorge, over which the 11th-13th Century ruins of Peveril Castle hang.[26]

The first cave that one enters is a gallery 340 feet (104 metres) long and 60 feet (18 metres) high. For centuries, until 1915, it was inhabited by up to 30 rope-plaiting troglodytes, who lived and toiled below ground, making ropes for the nearby lead mines, plus bell-pulls, and window sash cords. (They spoke a special language, devised in 1530, by one Cock Lorel, a resident bandit). Here underground were 'houses, barns for cattle, with stacks of hay and turf', with the candle-holding occupants doubling as guides to the cave system, as required.

More recently this great chamber has been used as a 'village hall', where film shows, concerts, Hallowe'en and fancy dress parties, along with ghost story sessions have been housed. Sounds of under-hill merriment, mixed with terror, may be heard in the vicinity.[27]

From the Great Chamber a passage leads to the Bell House. (Cotton reported that a bell that had been lowered into Eldon could not be pulled out. Perhaps it is housed here). Beyond that, one goes bent double, through Lumbago Walk, to Roger Rains House, where water cascades from its roof, thence via Pluto's Dining Room, and down The Devil's Staircase to the Half Way House, and the Inner Styx River[+]. Further still, the cave system extends for miles to link

+Note the Classical allusion; and the Romans, if not the Greeks, *were* here in the 2nd century.

with that from the Speedwell Cavern, but not, as far as now known, with Eldon Hole.

But if the goose that emerged in Castleton 'shell-naked, rifled of her plume',[28] has vanished, a *young* lady *does* emerge to roam on horseback, through Castleton's streets, on May 29th every year.[29] In doing so, she mingles 'History Time' with an annually returning present. She is dressed in a Stuart style of 1660, in honour of the return of Charles II to the throne, nine years after the Battle of Worcester, when he had hidden in an oak tree on May 29th. She is also personified in a small posy of flowers, known as The Queen,[30] which is placed on top of her male companion's bell-shaped flower bedecked cover, enabling him to literally ride *inside* her bounty. Together they process around the village, to celebrate the reviving year that they jointly embody. During this ride, they pause at each of six ale-houses, where village girls perform Morris dances, holding ribbon-decorated miniature 'Maypole' sticks, with music provided by the Castleton Brass Band.[31]

Eventually, the King's wire-framed heavy floral 'bell' is hoisted directly from his mounted position, onto the centre of the parish church tower, the other pinnacles of which are decorated with green boughs. The 'queen posy' is then placed on the village War Memorial, where the 'Last Post' and the National Anthem are played. Then a full-sized maypole in the market square becomes the focus for general dancing, to a lively 'criss-cross' tune.

In these ways, the community repeatedly interweaves the differing strands that contribute to its collective experience; the national with the local, shaft and tower, field and street, living and dead, male and female, Christian and Pagan elements, human and animal, underworld and summer skies, history into now-ness, as if the rope makers in the cave were still spinning an unbreakable sense of affirmation.

Overlooking, and rising less than two miles from Castleton, the crumbling remains of Mam Tor shape the Western horizon. 1,696 feet high, this Celtic 'Mother hill's name, has an affinity with Irish *mamm*, 'breast'.[32] The hill is now capped by a double banked and ditched Bronze Age ritual enclosure, constructed around 1200

B.C. It was probably the venue of an annual Lammas or 'new loaf' ceremony at the August 1st solar quarter-day, and start of harvest. With views over four counties, Mam Tor contains the 5-metre diameter platforms of 100 prehistoric houses, some with central hearths, post holes and abundant pottery, all probably of summer season use, rather than occupied throughout the year. Finds of a stone saddle quern, and a beehive rotating quern, prove that flour milling took place here. An unfailing natural spring and two round barrow burial mounds are other features of the hilltop. In Iron Age terms, this is the largest 'hill fort' in the area, though whether it ever served a military role is uncertain.[33]

As Fiennes noted, from the Castleton side, 'Mam Tor looks like a hay-ricke, cut in half', because the friable shale and thin sandstone beds have been almost continually trickling down, and have given the hill its other name, The Shivering Mountain.

The debris falls 'with a noise like thunder' says Moritz, to form another hillock named Little Mam Tor; yet, as has long been claimed, the Mother Hill does not shrink at all, despite giving birth to this 'offspring'.

Defoe treated this claim with scorn, since by his time England was strictly applying the method of scientific enquiry to all phenomena. Yet both the idea of a self-propagating Mother Hill, as a living, supervising entity, persists, as do her repeated landslips. From the Bronze Age to the present, the Hill has sustained both herself and her Little Mam Tor offspring, as the thousands who wander on and around her summit may see. As the exemplar of a regenerating landscape, Mam Tor epitomises the Wonders of the Peak.

The impossibility of clearly dividing 'practical' from 'mythic' affairs in Peak-land is demonstrated by the lead mine, believed to be of Roman, Saxon or Danish origin, that runs under the south flank of Mam Tor. It is now known as The Odin Mine, after the chief Anglo-Viking god.[34] Here miners found both lead and what they called 'fairies butter' or *gur*. Leaking from the rock, this milky liquor then evaporates, to form a honey–like substance. The word gur derives from German *guhr*, to ferment, and from Indian *gur*, the sugar of the gods.

Prior to modern times, in England all minerals were reputed to grow within the earth from 'petrific seed' and 'lapidifying juices', with fossils likewise produced from some 'plastick power of the earth'. And precious stones were believed to grow again, after their removal, while for modern geologists, limestone's billion fossils were all nurtured in a submarine habitat – presenting another wonder to contemplate.

V.
The Classical Legacy

ROMAN SILBURY

Roman Silbury and the Harvest Goddess

Fifty years after the birth of Christ, invading Romans first set eyes on Silbury Hill, Wiltshire, Europe's tallest prehistoric monument.[1] By then it was about 2,500 years old. Built using antler picks and ox bones, it stands as an achievement of the Neolithic or New Stone Age. The grass-covered truncated cone, 130 feet (40 metres) high, gripped the Roman imagination. Like modern passers-by, at first glance they also asked: 'What is it *for*?'

To answer the enigma we naturally turn to the scientific discipline of archaeology. Archaeologists have produced accurate measurements of the Hill's age, and the quantities and types of materials employed in its construction.[2] But as to Silbury's *meaning* they have remained notably silent, after confirming that it was *not* built as a tomb.

By contrast, the Romans seem to have recognised that to understand Silbury's mystery something more than ordinary calculation was required. The Hill arose from, and was permeated by, the sacred. In the old monument they saw a living, working goddess, rather than an inert thing. They came not to measure, but to *worship* her. By adopting beliefs passed down by oral tradition and seasonal behaviour from the era of the original builders, the newcomers accepted Silbury as a deity of profound importance to their own spiritual and material existence.[3]

Evidence for Roman veneration of Silbury, spanning their 350-year occupation of Britannia, lies very close to the Hill. Several Roman ritual shafts, rediscovered in the 19th Century, were dug only yards from the monument.[4] In addition, a Roman settlement, covering at least 22 hectares (54 acres)[5] has recently been detected around Silbury's northeast, east and southern flanks. This includes a stone-columned temple, 'an unusually important building', constructed due south of the Hill.[6]

Dr Bob Bewley, regional director of English Heritage South West from 2004 to 2007, believes that in venerating Silbury the Romans were 'drawn in the wake of their prehistoric forebears' by a sense of its divine value, expressed through pilgrimage to the monument.[7] If so, how may one account for such a remarkable continuity of devotion? What common issue or need could possibly have inspired both Neolithic and Roman worship at Silbury?

Perhaps the annual Harvest, concern for which they had in common? Agriculture was brought to Britain before 3000 B.C. Indeed it was partly Britain's renowned corn-growing reputation that attracted the Romans here. An ear of barley appears on gold coins from A.D. 10-40, inscribed with an image of the British king Cunobelin, while Roman coins of A.D. 45 showed an image of the corn goddess Ceres.[8]

As I suggest in *The Silbury Treasure* (1976) the desire to witness and participate in the rebirth of the divine 'corn child' probably inspired Silbury's construction, and might also explain its relevance to the Romans.

The excavations into Silbury's flat-topped cone, carried out between 1776 and 1968, in the hope of discovering a richly furnished Bronze Age burial beneath the massive mound, ended in disappointment. Recent scientific surveys of the Hill have confirmed that the monument is neither Bronze Age in date, nor a gigantic tomb, but a structure erected about 2500 B.C., by Britain's early farmers.

Yet the immense effort spent in doing so, using simple bone tools, renders the Hill an absurdity in practical farming terms, especially when agriculture itself was relatively new to Britain. Cultivation with crude wooden ploughs, hoeing, weeding, harvesting, winnowing and corn grinding by hand, were surely arduous enough occupations.[9]

Why should that hard-pressed community have erected a monumental Silbury folly? But that is to pose a question based on secular assumptions that were unthinkable in ancient times. Then, economic activity was seen as the *secondary* outcome of a primal sacred drama, in which the deities always came first. The deities were credited with repeatedly creating and destroying the universe, and all life forms, animal and vegetable. What divinities did, in the course of their cyclical life-stories, humanity tried to imitate. In this attempt myth-telling and visual symbols – both small and architectural in scale – combined with ritual performances.[10]

Therefore, to comprehend Neolithic and Roman Silbury we must look beneath modern 'common sense', and travel beyond exclusively 'objective' means of research, which often tend to cast an inappropriate if not obstructive veneer over the remote past. At least it is necessary to translate the data provided by technically orientated analytical researchers into something closer to the poetic norms typical of Antiquity. Then we may re-enter a lost state of mind, in which Silbury and her recurring First Fruits can be appreciated as gifts born from a divine Mother Earth, in collaboration with water, and the sun and moon's reflected power, lying on the body of her 'moat'.

Given such an outlook, both prehistoric communities and the Romans craved to worship the deity concerned. They did so in their fields, and around evocative rocks, rivers, or springs, especially during solar and lunar climaxes.

At Silbury they went further. They created a temple shaped like the Earth Mother's harvest-giving body. Thus Silbury's carefully modelled lake, from which the Hill arises, serves to describe a superhuman female effigy. To both Stone Age and Romano-British populations, Lake and Hill, when seen together, appeared as a 1,140-ft (347-metre) long image of the pregnant Goddess, in labour, over her sacred harvest child. The Hill, it is suggested, acted as her full womb. So, year after year, across the millennia, she displayed the profile of immanent bounty, to all who accepted her leading role in engendering the crops that ripened annually on her flat summit.

Just as the mammalian foetus is nourished from its mother's bloodstream, so Silbury Hill is composed largely of chalk quarried from the surrounding mother-shaped ditch, whose graceful, water-filled curves are modelled on a human prototype. Like the Silbury moat, the human body is 60 percent water-filled. Its form is reflected in miniature by innumerable Neolithic figurines

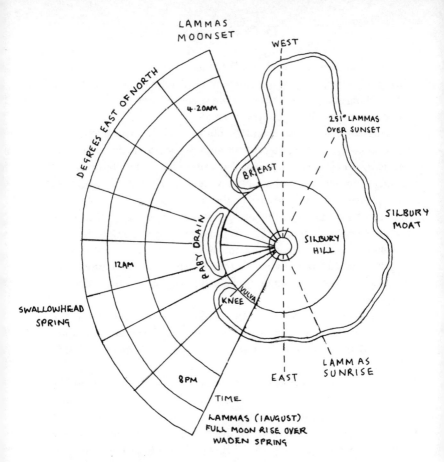

Silbury Hill showing the play of full moon light across its moat at Lammas, early August.

depicting the Goddess of agriculture, squatting like a human mother in 'childbirth'. As for the discrepancies in scale between these manifestations, a tiny crucifix around the neck stands for a god of human and cosmic dimension. Mythic symbols typically enjoy a flexibility of scale.

The incoming Romans (always notably willing to assimilate their deities with the native equivalents in countries they occupied), probably saw in the Silbury 'lake goddess', their own agricultural divinity, named Ceres.[11]

Ceres, along with her Greek equivalent Demeter, readily merged with ancient Silbury; hence a Romano-British pilgrimage town soon sprang up around the Neolithic mother hill.

As in human pregnancy, the material for the Neolithic chalk 'womb' (and stylised creation mound) was drawn from the 'quarry' that became the mother's still visible watery body.[12]

Two 'causeways' of undisturbed chalk, left *in situ* by the designers, define her 200 ft (61-metre)-long 'harvest child'. It was born when the full moon closest to August 1st (the solar quarter day, Christianised as Lammas), re-animated her water body from thigh to breast. Then the hilltop's sacred wheat could be cut as part of the First Fruits birth process, following which general harvest might start across prehistoric Britain. (See *The Silbury Treasure*, Chapters 4, 6, and 12).

Despite clogging of her moat by recent silt, the Silbury rock and water goddess remains visible for much of each year, more or less in her original form. Her curvaceous body can be seen by walking the periphery. (The Neolithic viewing terrace, just below the summit, is now out of bounds.)

Ceres in Roman Literature

Just as Roman art confirms a continuity of devotion to Silbury, the grandest expression of the eternal harvest Goddess in prehistoric Britain, so Roman writings help voice illiterate prehistory's reverence for that deity, whatever she was called.

Ovid's *Fasti* of the late 1st Century B.C., lists Tellus (Mother Earth) and Ceres festivals,[13] beginning in January, suggesting their joint adoration throughout the year: 'When the seed has been sown and the field fertilised... propitiate Earth and Ceres, the mothers of the corn, with their own spelt [a primitive mountain-wheat, also known to the Saxons], and flesh of teeming sow'. In Neolithic Europe and later, the Earth was often identified with the female pig. Pigs' bones were found in a Roman shaft near Silbury.

Ceres' name derives from the root *ker*, or 'growth'. Ovid prays to Ceres: 'O satisfy the eager husbandmen with boundless crops, that they may reap the due reward of their tillage. O grant unto the tender seeds unbroken increase... when the seed is buried, sprinkle it with water from the skies'. He begs the goddesses to ward off attacks by birds, ants, mildew, blight and weeds, before asking the farmers to repeat these prayers.

He ends by praising Ceres as a goddess of Peace, a role that she probably inherited from the relatively unwarlike New Stone Age: 'Under your foot long time war has been laid in chains... Peace is the nurse of Ceres, and Ceres is the foster-child of Peace', he affirms.

At the *Cerealia* festival, held annually on April 19th, Ceres was again honoured, for 'her bounty and services are manifest'. These 'Games of Ceres' involved offerings of spurting salt and incense [if available] and the waving of 'resinous torches'. In Rome in 484 B.C., the very first bronze sculpture ever cast was of Ceres, and placed in her temple near the Aventine, where the grain ships moored. As in rural Britannia, it was the common people (the Plebs of Rome) who were especially devoted to her.

In his book *Metamorphoses* (the source for several Romano-British mosaic designs), Ovid describes Ceres as 'that most beautiful goddess'. By 'nodding her head she made the fields laden with corn to tremble'. Destiny, he adds, 'does not allow Ceres and Hunger to meet'.[14]

To Romans, Ceres embodied the Neolithic origin of agriculture. She was therefore the Silbury Goddess' equal. Ovid writes: 'Ceres was the first to break up the sods of earth with the crooked plough; she first planted corn and cultivated crops. She imposed the first laws in the world. All we have, we owe to Ceres. Of her must I sing'.[15]

He describes how, in Italy, on the verge of Harvest, 'all married women celebrated the annual festival of Ceres... Dressed in garments of snowy white, they offer her garlands of wheat ears, the first fruits of the crop'.

Thus they *participated* in Ceres' annual rebirth of her sacred corn child. For nine nights they avoided physical contact with men, in order to share without distraction the corn goddess' labour pains.[16]

Perhaps the same need for ritual empathy explains why 'fourteen Romano-British buildings face the mound', and why more than two hundred Roman coins, presumed to be votive offerings, have been collected from Silbury's ditch. Two platforms, one cut into Silbury's northern face, the other into its southeast side, may have been used to hold Roman altars, according to the archaeologist David Field, who spent years carrying out a survey of the hill.[17] The south-eastern platform overlooks the place where the annual 'corn child's' head emerges at Lammas, from the 'hill-womb',[18] assisted by the full harvest moon, combining to make a vital focus for pilgrims.

In Virgil's *Aeneid*, (composed in the late 1st Century B.C.), Ceres plays a key role in Rome's birth. On fleeing a burning Troy, the demi-god Aeneas gathers fugitives together on 'a hillock containing an ancient and deserted shrine of Ceres'.[19] When eventually he and his followers, as prophesied, make landfall near the Tiber's mouth, they sit down to eat on a 'flooring provided by Ceres', namely grass; (proto corn). The only 'plates' at this first feast are Ceres' 'meal cakes, thin cereal platters... the round-shaped crust of destiny', which they eat. Thus before Aeneas can create Rome the goddess of agriculture is recognised as vital to its survival. All cities depend on Ceres' food.

Due to the warmer climate, Ceres' wheat in *Italy* was ready to be cut by June 15th, when the *penus Vestae*, or symbolic grain store of the Roman state, had been purified to receive the sacred First Fruits. Then on August 25th the Latin goddess Ops Consiva, 'Abundance personified', was celebrated.[20] She presided over the fully garnered crops.

Tellus and Ceres

Ceres and Earth (alias Tellus), are 'Partners in labour, and they discharge a common function', writes Ovid.[21] Thus the April Cerealia festival was paired with, and preceded by, the April 15th Fordicidia festival of Tellus, at which a pregnant cow was sacrificed. Tellus was the 'ground mother' in whom Ceres, the mistress of farming processes, operated. Tellus, whose name derives from Latin *terra*, 'earth', was the foundation, sustaining every sacred imagining, a deity set to outlast all human constructs. She may be seen around Silbury in the neighbouring hills and streams. These natural features were revered by Saxon, Roman and Stone Age peoples as aspects of a living sacred Earth.

The Myth of Demeter and Persephone

From Greece, the Romans brought to Britain the myth of the mother goddess Demeter, alias Ceres, and her daughter Persephone, alias Proserpina. Their story closely matches the divine annual drama described by, and enacted at Silbury and the other Neolithic Avebury monuments.[22]

Roman Myth and Neolithic Monument

Proserpina, also known by the Greeks as Kore, 'The Maiden', is the epitome of Spring. She and her sea nymph companions go picking flowers in a meadow. But when she plucks a narcissus the ground gives way beneath her. She is dragged into the abyss by Hades, ruler of the underworld. Down there she is obliged to marry him, and becomes Queen of the Dead; whereupon her Greek name changes from Kore to Persephone, meaning 'to bring or cause death'.

Ceres (Demeter) is distraught at the loss of her daughter. A 6th Century B.C. Homeric Hymn[23] describes how 'a sharp pain' seized the mother goddess's heart. Using two pine trees as torches (adds Ovid) she searches for Proserpina by night and day, through air and over ocean, with a pair of serpents to pull her chariot, yet fails to find her daughter.

Embittered, Ceres decrees that there will be perpetual famine on Earth until the lost child, her 'sweet sprig', is returned. But Proserpina has *married* Hades, and cannot desert him. Then Jupiter (Greek Zeus) decides that she shall come back for eight months of every year, and only spend each winter with Hades. So there *will* be another harvest, sprouting from the underworld to the surface.

Then a young demi-god (called Triptolemus in the Greek version of the myth), is instructed by the mother goddess in the art of agriculture, which he carries around the world. Thus, through their sacred drama, the deities recreate each season *and* the prototypical agricultural round, as experienced and followed by humanity.

In essence, Ceres' myth is as old as the Neolithic birth of agriculture. It enabled the Romans to participate in the annually restaged cyclical drama, act by act, at the seasonally appropriate Neolithic monuments, arranged around the Swallowhead springs. So

Romans offered coins at the West Kennet Long Barrow, one of several potential 'underworlds' available to them in the vicinity. For them, as for native British farmers, the similarities with Classical myth were clear. Therefore the newcomers honoured the Neolithic framework and built their pilgrimage settlement at Silbury's foot.

In Roman, as in Neolithic myth, the Daughter-Mother-Death hag modes of the goddess were essentially aspects of one *deity*, just as the four elements and seasons make a dynamic, unified whole. Thus the Silbury moat length fits *exactly* across Avebury's henge arena, while each noonday sunbeam links the three major monuments and Swallowhead springs together in a repeatable drama.

Silbury's Roman Shafts

In the search for renewal, the Romans needed direct communion with underworld powers. Perhaps that explains why they dug four vertical shafts, up to 26 feet deep, and each less than 120 yards (110 metres) from Silbury's life-giving flanks. Archaeologists now accept that 'a religious function was served by these shafts'.[24]

In the 19th Century, a 'Mr Kemm of Avebury' discovered one shaft sunk 100 yards (91 metres) from the 'thigh' of the moat image, aligning with the start of the annual Lammas birth event. The other shafts, set in an arc, and similarly close to the hill, are perhaps intended to follow the sacred parturition, to its fulfilment at the 'moat child', as defined by the two causeways. If so, they serve to emphasise that the moonlit monumental mother and child also depended on the hidden underworld for their successful renewal.

Although the shafts strike water on encountering the clay layer that gives rise to Swallowhead and the Silbury moat, these are no ordinary wells. Indeed, with so much pure surface water nearby, they are unnecessary in prosaic terms. Two of the shafts have been excavated. The Romano-British material found therein includes beads, glass, a bronze ring, dozens of coins (some of early 5th Century date), an iron meat hook, shears, and a stylus, along with abundant pottery including jug handles, and a 12-in (30-cm)-high stone 'phallic' pillar. All these artefacts were mixed with bones of horse, stag, pig and at least one human infant, arranged in underground communion.

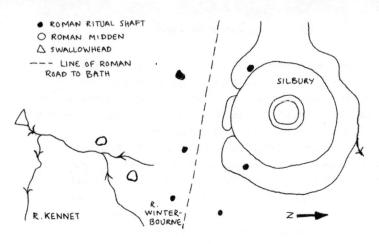

Silbury and its relation to the source of the River Kennet.

The material seems to be a deposit, offered to indigenous spirits of place *and* to the imported yet compatible Classical deities of the agrarian cycle. It is unclear whether traces of large Roman granaries detected half a mile northeast of Silbury were for locally produced grain or for stored tithes offered to the Silbury Goddess from afar – like those brought to Demeter's Eleusis shrine from all over Greece.

Further evidence of continuity with Neolithic harvest intent is given by the Roman sickle hone or sharpening stone, found inserted into Silbury's east side, on an ash-strewn ledge, assumed to have been cut in Roman times. In addition, the Silbury-Swallowhead connection was confirmed in 1867 by the discovery of a 12 ft (3.6 metre) long Roman pit, sunk midway between Silbury and that spring. The trench contained the debris of ritual feasts, including corn-grinding *mortaria*, a fine drinking cup of Castor Ware, oyster shells and coal, together with animal and human bones.[25]

Roman devotion to the Harvest goddess Ceres was often extended into the imperial cult, as shown by the deified empress Julia Domna, Syrian-born wife of the emperor Septimius Severus. She accompanied him to Britain in A.D. 208, and is described in a Romano-British inscription, found near Hadrian's Wall, as: 'bearer of corn… Mother of the Gods, Peace, Virtue, Ceres… weighing life and laws in her balance.'

In summer zodiac terms, the donor adds that Julia-as-Ceres is also:

'The Virgin in her heavenly place [who] rides upon the Lion'; i.e. late August mounted on early August. Therefore she personified both the divine First Fruits birth and the ingathering of general harvest. Like Silbury, Julia claimed to bring abundance to all. However, unlike Silbury, her immortal power to do so ended with her suicide.

Silbury and the Maternal Juno

It would be wrong to regard Roman interest in Silbury as merely a matter of securing bountiful corn supplies. Given that the Romans arrived with deeply entrenched devotion to Jupiter's sister, Juno, they were well equipped to replicate, with Juno's help, the corn-child, *double* birth act known to the Neolithic Silbury builders.

Juno, the very great matronly goddess in whose honour our month of June is still named, had two ancient titles – Lucetia, and Lucina. As Lucetia, she was the feminine principal of celestial light, including that of the moon, of which she was the goddess. Therefore she was perfectly equipped to annually re-enact the full moonlit rebirth of the harvest child from the Silbury moat, in the Neolithic manner, with water body now seen as her own.

As Juno Lucetia, her sphere of influence clearly included the sky, so that she was able to sustain the sky-earth link envisaged by the monument's designers. She was guardian of the young bride, and attendant for women in labour. In Rome, her temple was built only a few years after the city's foundation, in 735 B.C., while in another Roman temple, she was worshipped as Protectress of all the Roman people. Since her cult was Empire-wide, it seems highly likely that she was instrumental in translating Neolithic Silbury to meet the Roman norm that it almost perfectly matched.[26]

By working together Ceres and Juno were able to stay true to Silbury's original purposes.

So, as Juno Lucina, the Goddess of childbirth, she delivered her new-born baby into the light, thus wrapping together the two aspects of her power. Moreover, under her title of Caprotina, she was also the goddess of Fertility, thus ensuring a secure link between human and agrarian fecundity.

161

RECOVERING BRITAIN'S
CLASSICAL GODDESSES

History declares that each moment in time, once gone, is lost forever, and that we are doomed to live within an ephemeral state, largely disconnected from past and future. By contrast, the mythic outlook enjoys time as a cyclical event, comparable to the recurring seasons, and the regular movements of stars, sun and moon.[1]

When, as now, the historicist attitude dominates, it produces a frenzied desire to somehow keep in touch with lost epochs, by unearthing artefacts, usually assessed in contemporary terms, which therefore largely fail to mitigate the loneliness of excavators, operating within an isolated present tense.

Yet their efforts are generally welcomed, including by those who are inclined towards a mythic outlook, since for them, these recovered images speak of repeatable truths, delivered in a variety of unfamiliar accents, which collectively help to enrich and confirm the nature of recurring time.

Yet to encounter a discarded deity, such as the gilded bronze head of the Goddess Minerva (opposite), found in a corner of the Roman Baths at Aquae Sulis,[2] is shocking – both because her head has been so violently severed from her lost bronze body, and because she continues to speak silently, through parted lips, across the millennia, to many generations.

As a principal member of the Graeco-Roman pantheon, no fewer than 18 inscriptions carved in Minerva's honour have survived in Britain. She derives from a combination of the Etruscan thunderbolt deity, combined with the Greek Athene. Her name is associated with the Latin word *mens*, 'mind, understanding, feelings, and heart'.[3] In ancient Rome, her chief festival was held at the spring equinox, and it lasted for five days.[4] Minerva was especially revered by artists, musicians, and doctors.[5] At Bath, in a typically Roman manner, she attached herself to the presiding British goddess, Sulis.[6]

Britain in its entirety had probably been regarded as a living, divine being, by its Celtic and pre-Celtic inhabitants. For example, Ireland, (as

Gilded bronze head of Minerva, 10 inches high.
2nd Century A.D., found in Aquae Sulis (Bath).

the *Eriu* derivation of her name makes clear), was the Goddess. The Celtic and pre-Celtic inhabitants' view was further supported by the incoming Romans, through their visual arts, inscriptions and coinage.

Thus Marcus Cocceius Firmus, a Centurion of the Second Legion Augusta, set up a gritstone altar, dedicated to 'The Genius (or spirit) of the land of Britain'.[7] Similarly, carved onto the base of a Britannia statue in York, the text 'To Holy Britannia', is still legible.[8] Her divine life-force, reaching from coast to coast, was supported by numerous dedications to *genius locii*, the 'spirits of place' attributed to particular localities, of which 18 examples have survived.[9]

In A.D. 117, the Emperor Hadrian introduced onto our coinage an image of a water-surrounded Britannia, armed with spear and shield and enthroned on rocks.[10] When Hadrian decided to rename Albion as the female Britannia, he was supporting the continuity of a long-established island-as-Goddess tradition.

Britannia's image was revived by Charles II, in 1672, and she continued to feature on our currency until 2008.

Roman goddesses could also merge into a deified Empress, as demonstrated by inscriptions found about Julia Domna, the wife of Septimius Severus, who ruled from A.D. 193 until 211.[11] On a panel

163

found on Hadrian's Wall, this 'Empress-Goddess' is praised as the 'bearer of commerce, inventor of law, and founder of cities, by whose gift it is man's good lot to know the gods'. Therefore she is Mother of the Gods, and of Peace, Virtue, and of Ceres. She is also Astarte, the Syrian Goddess, weighing life in the balance.[12]

The range of deities to whom written dedications were made in Britain reflects the influence of Middle Eastern and Europe-wide culture and mythology on the Roman Empire. Serving soldiers were posted in Britain, bringing their beliefs with them. Romans were also willing to absorb Britain's own pre-Roman goddesses into their capacious religious outlook.[13]

For example, the great popularity of the Mother Goddesses in Roman Britannia was helped by their affinity with the indigenous Matres[14], in which a row of three carved 'Mothers' hints at their alternative 'maidenly' and 'crone' forms, hidden within their maternal manifestation.

A Chichester inscription refers to the Matres Domestica,[15] yet they become Matres Sulevis in Cumbria[16], and are identified with The Fates[17] in several other instances. A number of examples explicitly link a dedication to the Mothers with those of the serving soldier's overseas homeland. He prays to her for his sustenance and survival.[18] Likewise, the Campestres,[19] the Mother deities of the Parade Ground, have five surviving inscriptions. Military drill took place under her protective regard. Belief in a universal Great Goddess may underlie this whole group of appeals.

Combining agriculture with politics, the divine Salus, alias 'The Queen',[20] has five British altars, reflecting her substantial worship in Rome, where her birthday feast was held on August 1st. As 'Salvation' personified, she stood for the welfare of the State. She protected crop-sowing, and was depicted on coins holding ripe corn ears.

Fortuna, otherwise called Bonus Eventus,[21] has 32 British dedications. As Fortuna Conservatrix she is invoked, 'following a vow, made at a time of danger,'[22] by Marcus Aurelius Salvius, tribune of the First Cohort of Spaniards.

Britain contains evidence of devotion to the Egyptian Goddess Isis, who had a temple in London, which was restored in the 3rd

Century.[23] Perhaps less exotic, but just as important, a bronze effigy of Ceres, the Roman goddess of the corn harvest, was found in the London Thames. She is shown holding a bunch of corn ears.[24] More concerned with wilderness than farming, the huntress-lunar goddess, Diana, has six surviving references in Britain.[25] Her temple at Caerleon, in South Wales, was restored in the mid-3rd Century by Flavius Postumious.

With less surviving evidence of her popularity here, Nemetona, a goddess of the sacred grove,[26] has only one altar in Britain. It was set up by Perugrinus from Trier and was found in Bath. Likewise, only a single altar has been found honouring Bona Dea, 'The Good Goddess', whose worship was restricted to women only.[27] Despite her name, Abundantia, is depicted in Britain by only one surviving effigy, found at Lydney's temple of Nodens.[28] Far-reaching Aeternae, 'Eternity', alias Roma, whose task was to protect the entire Empire forever, has an altar and votive pillar at Maryport, Cumbria.[29]

An inscription, written in Greek, to the Syrian Goddess, Astarte,[30] was found at Corbridge, Northumberland. It declares: 'Pulcher set me up.' For him, this altar was a 'me', not an 'it': and the job of this 'living stone' was to tap and to concentrate some of the earth's underlying vitality. In this effort, the cylindrical volutes, that flank the altar-top focus (where the flaming offerings were made), acted as 'intelligent eyes', to help guide and establish the difficult connection.

Elsewhere, the goddess Nemesis, sometimes termed Lady Nemesis, was summoned, as when, 'following a vision', she was called up by Marcianus of Chester.[31] Nemesis was feared by those who had offended against the moral law, or had accumulated excessive wealth, or had displayed arrogance before the gods, for she was the mistress of equilibrium. In Britain, one supplicant, in recompense for his errors, promises 'to give her a cloak and a pair of boots'.

Thetys, wife of Oceanus,[32] has one known dedication on this island, as has Juno,[33] consort of Jupiter, (though he has dozens). Dea Tacitus, consort of the dead, is recognised only once,[34] but Hygaeia, the Goddess of good health, is recorded in two places.[35]

Epona, the imported Horse Goddess, is recognised with two

altars,[36] while the local Goddess Brigantia, the 'High One'[37] of a powerful North Country tribe, is twice acknowledged as Victoria, or as Caelestis, (Heavenly), and in South Shields as a 'Divine Nymph'. Her fine, classically attired statue, found at Birrens, Dumfriesshire, shows her holding an orb and a spear.

Among the eight dedications to The Nymphs, Chester's Twentieth Legion set one up, while at Greta Bridge, Brica and her daughter Januaria, were responsible for another.[38] At Risdale close to Hadrian's Wall, an anonymous soldier, 'for-warned by a dream', 'bade her who is to marry Fabius to set up this altar', 'dedicated to the nymphs who are to be worshipped'. Her wedding present took the form of a duty.

As beautiful, youthful deities of wells and springs, The Naiads[39] can also be seen reclining on the margins of 4th Century mosaic pavements at Woodchester, Gloucestershire, and at Rudston, East Riding. (The latter is now displayed in Hull Museum). On altars found elsewhere, the Naiads are said to be supervised by a goddess similar to Arnometiae,[40] who presided over the holy well at Buxton, Derbyshire. Moreover, all British rivers were the outward expression of a water deity, such as Sabrina of the Severn and Verbeia,[41] whose inscribed altar was found on her river bank at Ilkley, Yorkshire. There was clearly a strong topographic element behind the Roman impulse to worship. Nevertheless, a significant number of dedications make a general appeal to 'All the gods and goddesses'.[42]

Of sacred images related to Lararia, Roman household shrines, the most frequently found are those of figurines of Venus. Her figurines, made of white paste, were mass-produced and imported from Gaul. Of these, hundreds have survived, either whole, or in part.[43]

Of other female deities who have left their traces in Britain, there is the snake wrapped Gorgon's head, and several obscure divinities such as Garmangabis, a Germanic 'Great, Generous Giver', commemorated at Lanchester, Co. Durham.[44] We also have effigies of the war-loving Bellona.[45] As a sister of Mars, her 'priests' were Gladiators. By contrast, Concordia[46], whose name appears on the base of a low relief carving, epitomises harmony, as does Pax (Peace), whose altar was erected by a tribune, serving as a prefect in northern Britain. From the sublime to the best hopes of traders, Latis was a lake deity, worshipped by brewers of beer and mead, especially around Burgh-by-Sands, Cumbria.[47] In deeper waters, Panthea had underworld powers. She was a sorceress, revered by sailors.[48]

Back on land, Ratis was a deity who protected ramparts and forts[49], where The Three Witches, (the Lamia Triades) also featured, as a carving from Benwell, Cumbria, makes clear.[50]

The nine Greek Muses, from the foot of Mount Olympus, were depicted on a mosaic pavement at Aldborough, Yorkshire. Of these only Thalia, the Muse of Pantomime, remains visible, though she still carries her traditional 'grumpy old man mask'.[51] Among lesser known deities, Setlocenia,[52] 'the nurse of warriors', was worshipped by Labarius, a German; while at Catterick, Harimella was adored by an engineer named Gamudiahus.[53] The goddess Hammia's devotees were restricted to one cohort of the Syrian archers, based at Carvoran.[54] Also limited in her appeal, the Goddess Ricagambeda was worshipped only by one section of the 2nd cohort of Tungrian soldiers,[55] while the goddess Viradecthis was associated with another part of the same unit.[56]

Most of Britain's altars and inscriptions were paid for and erected by men, for whom the Goddesses addressed every aspect of their lives, just as they did for women In their joint understanding of a divinely articulated world, feminine attributes were eagerly acknowledged by most people. But whether we regard any of these unearthed relics, now visible in our museums, (which serve as our Temples of the Muses), as valuable components in our *present* cultural landscape, is a matter for individual judgement. At the very least they offer an antidote to imprisonment within isolated self-hood. In addition, we can presumably share their pleasure in the seasons,[57] as depicted on mosaic floors, where they parade, garlanded in flowers, grain crops or grapes, or cloaked against the cold, according to the time of year. In doing so, they offer us their collective gift of a shared timelessness, that belongs to all humanity. They remind us of this precious legacy. We can find these women, set into the four corners of mosaics preserved in Cirencester and at Bignor in Sussex, among many other locations.[58] On a mosaic at Littleton, Berkshire, the seasons are depicted in a circle of four galloping animals, on whose backs ride the Mother Goddess

Demeter, and her daughter Persephone, shown waving 'Goodbye', as she plunges into Hades' underworld.[59]

We may also have good reasons to notice the city-protecting crowned female Tyche figures,[60] who feature on some mosaics, such as that from Brantingham, now in Hull Museum. Whether concerned with the well-being of our towns or rural areas, these inherited effigies can help to focus our efforts, as we work to secure Britain's present and future health.

Considered overall, the Classical female deities rediscovered in Britain, accumulated during nearly four centuries of Roman occupation, can still bring us into a substantial relationship with a divine 'otherness', so giving to 'ordinary life' a richness that invests the intimate with the infinite. Even when defined as abandoned 'fictions', that is what the old deities may continue to do. Just as we retain an appreciation of Classical literature, as a reservoir able to slake the parched imagination, a pool bobbing with half-submerged truths, so, too, our endowment of Classical deities can reconnect us to a fuller, less parochially mechanised world.

And of course Classical deities continue to underlie our two early summer months, May and June, which are named after the Greco-Roman goddesses, namely the earth divinity Maia, mother of Hermes, and Juno, sister of Jupiter, respectively. Together they cover our islands with burgeoning flowers and grasses, and every tree's foliage. We can surely enjoy their company.

WRITTEN IN STONE

Our understanding of a previous civilisation's attitudes may be greatly illuminated if their ideas are found carved into stone, especially when text and sculpted image occur together.

The famous hot springs of Bath were named Aquae Sulis, and several commemorative stone altars are dedicated to the British goddess of the hot springs, named Sulis. The waters of 'Sulis' had been adopted by the Romans, in the 1st Century A.D. Sulis was the

Three Roman altars, inscribed with dedications to the goddess Coventina,
found near her spring, close to Hadrian's Wall.

deity who was identified with the source of those waters.[1] They emerge from the earth's interior, through a fault in the limestone rock, at the rate of 2.3 million litres a day, and at a temperature of 45 degrees Centigrade.[2]

Her name, *Sulis*, derives from proto-Celtic *suli*, 'the sun', and Celtic *sùil*, the eye. Here the sun in the sky combined with the 'eye' of the spring, to render it hot. Sulis was also known as the 'Bright One', and a fire was tended perpetually in her temple on the site. In her, the elements of earth, fire and water combined, to present an elemental union.[3] (And in Celtic myth the sun is usually regarded as feminine.[4]) By 60 A.D., the Romans had driven oak piles around the muddy spring-head, and built a lead-lined stone vault over it, which has survived to this day.

The incoming Romans apparently accepted the Celtic version of elemental synthesis, and retained Sulis' name, placing it *before* that of their own goddess, Minerva, (who promoted the arts, including music and medicine), to create Sulis-Minerva, as the double deity, worshipped at Aquae Sulis. Yet it is Sulis alone who is invoked on most of Bath's commemorative stones. Many of these are altars, upon the centre focus of which, a fire could be lit. Six have been found at Bath,[5] and others, sometimes addressed to the Suleviae, (a triple form of the same deity) have been unearthed at Cirencester[6] and at other places in Britannia. On these stones we are given the names of the person remembered and that of the dedicator, who usually 'willingly fulfilled his vow'[7] in carrying out the commission.

Sulis was usually imagined as a matronly figure, associated with fertility and child-bearing, as testified by the bronze and ivory breasts deposited around her spring. Her waters were also credited with curing rheumatism and gout, and with recovering lost property. Thus 130 'curse' tablets were addressed to her, some written on pewter sheets, and in code, begging for her help in tracing items. For instance, Docimedes, whose pair of gloves had vanished, asks that the thief should 'lose his eyes and mind in the goddess' temple'.[8] Over the entrance to that temple, the circular image of a winged, snake-surrounded, gorgon's head stares down. Yet 'She' is given a moustached masculine face, perhaps to match the male sun, Sol, of Roman belief.

On Hadrian's Wall, a local deity named Coventina was celebrated in carved imagery and by several altars, dedicated to her, and ten inscriptions bearing her name. This notable collection of objects comes from her well at Procolitia, Northumberland, excavated in 1876. The seven ft (2.1 metre) square stone-lined well, with a clay-smeared inner wall surface, was first rediscovered in 1732, and its waters are still flowing.[9] The well was surrounded by a square enclosure opening to the west, towards a priest's house. That the term Augusta was applied to Coventina, on two ceramic incense censers, suggests she was very highly regarded.[10] Again, as in Bath, these vessels establish that fire was an important element in her worship. She was a deity of abundance, inspiration and prophecy.

The well was probably built by Roman engineers c .A.D. 130, to help drain a marshy valley, but this structure took on religious meaning thanks to the Celtic inhabitants of a nearby settlement. Her cult was at its height in the 2nd and early 3rd Centuries, but in A.D. 391 an Imperial Edict closed down all pagan temples, whereupon her altars were carefully laid in the well.[11]

Coventina's name may derive from Celtic gover, 'head of a rivulet',[12] or from the Celtic word, *cof*, (pronounced 'v'), meaning 'memory, or remembrance'. Votive pots, some bearing her image, and a huge mass of copper coins were found in the well, along with boar's tusks, deer horns, five bracelets, and bronze images of a horse and a dog.[13] One of the eight altars dedicated to her also carries an image of her head, while another shows her in a niche, reclining on a lily leaf, holding a frond of vegetation in her right hand, while pouring water from a vessel. On other altars she is referred to as a Goddess, but sometimes as a Nymph, or a Goddess-Nymph. A splendid three ft. (91 cm) wide stone panel, depicting three water nymphs, seated beneath an arcade, each holding a beaker in one raised hand, while pouring water from urns, emphasises the triple-nymph aspect of this deity.[14]

At nearby Carrawburgh, the remains of another Shrine to the Nymphs has been found, consisting of a well, a semi-circular stone seat, and an altar pedestal, dedicated to 'the nymphs and genius of this place', by Marcus Hispanus Modestinus, Prefect of the 1st cohort of Batavians. He dedicated it 'willingly and deservedly on behalf of himself and his family'. The inscription is accompanied by carved images of an ewer, a patera, a strainer, and a cleaver.[15]

The legacy of carvings and inscriptions, accumulated during the centuries of Roman occupation, is strongest when invoking the *Matres*, or triple mother goddesses.[16] Their cult was probably Celtic in origin, and was well established here prior to the Romans' arrival, yet it was *they* who gave the cult more than 50 triple-figured surviving carved images, and dedicatory plaques. 'The mothers' are typically shown seated, carrying both fruit and loaves on their laps. Sometimes they are accompanied by young children.[17] The mood that they convey is one of contentment. Their popularity lasted deep into the Middle Ages, for they addressed issues of a timeless human concern.

At Castlesteads, in Cumbria, a centurion named Gaius Julius Capitianus 'restored a Mother Goddess temple, fallen through age', says his surviving inscription,[18] which is now in Carlisle museum. His deed, recalled by his rock-carved words, is literally *grounded* in our *common* humanity, and into the physical fabric of this island. That the Matres were never individually named, contributes towards their wide scope.

Today, from end to end of England, their inscriptions and surviving carved images, nearly 2,000 years old, can join with the remnants of our folk tradition, to sound together in an improvised chorus, that honours and celebrates the ages-old goddesses of continuity. And as Sulis revealed, fruitful matrimony rests upon a marriage of the elements, and shares in their magnificent array.

AN EFFIGY OF SABRINA

In depicting every stream and river in England as inhabited by a female water spirit [1], Michael Drayton was following a long established convention, prehistoric in origin, in which all running water was regarded as a living, sacred gift.

Sabrina was the living embodiment of Britain's longest river, the Severn, as recorded by Tacitus in the 1st Century A.D.[2] 'Severn' is a corruption of Sabrina's name, which, according to Ekwall's *English River Names*, derives from the ancient Sanskrit word *saba*, meaning 'milk', an appropriate link, considering the fertile pastures of her valley.

In a fable, recorded in Geoffrey of Monmouth's *History of the Kings of Britain*, c. 1130,[3] Sabrina is equated with the beautiful, illegitimate daughter of King Locrinus, by his mistress Estrildis, who he had incarcerated in a cellar beneath London for seven years. When his wife Gwendolen heard of this she raised a Cornish army, and defeated Locrinus at or near Stourbridge. She threw Sabrina into the Severn, where she has lived on, ever since. A bronze statue, depicting this act, and made in 1880, stands in the café foyer of Worcester's city art gallery.

In 1637, John Milton's masque *Comus* celebrated Sabrina's continuing life beneath the waves, from which she rises, accompanied by her water-nymphs, to briefly sing and dance on the riverbank.[4]

Between 1770 and 1995, the lower part of Gheluvelt Park served

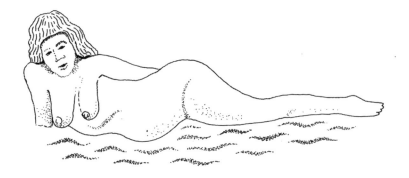

Sabrina, goddess of the river Severn;
she was recognised before, during and after Roman times.

as the City's reservoir, with water drawn directly from the Severn, initially by a wheel, and eventually by pumps, feeding sand and gravel filter beds covering 7.5 acres. Wooden, and later, cast iron pipes fed the supply to almost every Worcester house.

Geographical Setting

From Pumlumon ('five beacons'), a mountain in mid Wales nearly 2,000 feet above sea level, Sabrina-Severn runs for 220 miles (354 km) into the Bristol Channel and the Atlantic. The river is tidal as far upstream as Gloucester.

Prior to the last Ice Age, her headwaters drained north into the Dee at Chester, but when ice blocked that exit, she cut a gorge through the ridge at Ironbridge, to enter, merge with, and enlarge the Stour valley.[5]

Sabrina frequently re-enters the lower end of her former Worcester waterworks, and runs around the nearby race course during most winters.

Towns situated along the river include Llanidloes, Newtown, Welshpool, Shrewsbury, Ironbridge, Bewdley, Stourport, Worcester, Tewkesbury, Gloucester, Avonmouth, Weston-Super-Mare, Minehead and Ilfracombe.

In 2013 I put forward a proposal to site an image of Sabrina in the former water-works at Gheluvelt Park, Worcester, in a drawing together of numerous historical and geographical references. It would complement the World War I memorial placed at the other end of the Park.

Depicted in highly durable fibreglass, containing a metal armature, with prongs set into a concrete base, this enduring, majestic, effigy was intended to celebrate the magnificence of Nature's local endowment – including the vegetable and animal life, fed by the river – in a superhuman, double life-sized form. It would have offered an extra way to connect an urban population to the wonders of its natural and cultural environment.

Inlayed around the base of the reclining figure, a miniature version of the 'river' was to be depicted by inlayed ceramic and glass tiles, with the riverside towns represented by red glass spots.

By drawing different *kinds* of knowledge into a single working image this artwork would have created an enlarged *type* of understanding.

My proposal was considered and rejected by Worcester City Council in February, 2014, on the grounds that the style of the figure was too voluptuous and not in keeping with the park environment, and that it could attract anti-social behaviour.

But the voluptuous Sabrina continues to make her presence felt. Only a week after my plan for a Sabrina effigy was rejected, the river burst her banks and flooded across the entire park, which was consequently closed to the public for ten days.

V.
Horse Goddesses

THE UFFINGTON HORSE DEITY, INCLUDING THE 'CASEG FEDI', AND THE 'MARI LLWYD'

From the escarpment's edge, the Uffington White Horse hill figure gleams over the 20-mile-long, fertile Vale which carries its name.[1] This horse has attracted numerous accounts written from the 12th Century onwards, when it featured as the fifth Marvel of Britain in a list compiled by Ralf de Deceto.[2] It was the chief subject of Thomas Hughes' book of 1859. Now known to date from the prehistoric Bronze Age, c. 1000 B.C., the horse is depicted galloping, as if permanently on the move; with an overall length measured at 365 feet (111 metres).

Where is the Uffington White Horse going? In the following account, I try to show that the Uffington White Horse's dynamic pose is fully justified, for she is engaged in a busy, repeatable round of activity. Secondly, I believe that the horse image was specifically sited in response to at least three striking natural features, found in its immediate vicinity.

As the fastest animal in Britain, horses frequently carry solar references, running from sunrise to sunset,[3] a task emphasised by the east to west positioning of this white image, using the underlying chalk rock, to match the sun's dazzling daily and annual rides across the sky. Similar 'sun' horses are well known in Irish and Welsh myth and legend,[4] and on early pre-Roman British coins, where the beast is often shown together with a wheel, another solar emblem, chosen to epitomise movement, as on the gold Dobunnic West Country coins of the 1st Century A.D.,[5] where the two symbols are combined.

Folk behaviour typically re-enacts pre-Christian practice, and here, as recently as the mid-19th Century A.D. it was customary to roll a wheel from the adjacent 600-metre-long steep comb, now called the Manger,[6] towards the source of the River Ock at its foot. This was done in springtime and at midsummer, when the Horse's position, perched high on the escarpment at 245 metres, evokes the sun's combined noon-day and mid-summer apexes. In some British accounts, the launched wheel is described as wrapped in flaming straw, to further emphasise the solar link.[7]

Hughes says that the Manger was so called because that topographical 'great hole' lies right under the horse's nose, to serve as its 'hay bin', and adds that the Manger's fluted west flank was called the 'Giant's Seats'.[8] Around Uffington we may apparently have encountered giants as eager spectators of a superhuman event.

If the term 'Manger' suggests future nutriment for the horse, this valley's earlier title, Hrungrpyt,[9] found in 10th Century Anglo-Saxon estate charters, hints at a seasonal dynamic. The Saxon word *Hrungr*, or *Hring* meant: 'a ring, circle, or circuit', while *Pytt*, was 'a hole in the ground, a grave'.[10] In Hrungrpyt, the circling year involved celebrating an annual marriage of extremes, in a reunion of the elements.

This union included the notably spectacular cluster of the river Ock's springs, hidden in the forested gully, among a steep jumble of fallen and new trees, in a continuation of the Hrungrpyt's gulley, while the sky-high and bone dry horse badly needed to drink at the Ock's source.

The O.E. word *Aesc*, 'ash tree', was formerly applied to the crest of the Downs, hereabouts,[11] and the Ash was the sacred tree and vertical axis of Anglo-Saxon mythology. Here it helped to join underworld, sky and earth's surface together, a union that required a festival in response, which combined ritual trade and sports.

Hughes' account of 1859 includes details of the 1776 Uffington gathering, attended by 30,000 people, at which there were races for horses, cart horses, and asses, a sack race for men, (who were asked to bring their own sack), and for the women's race winner, the prize of a smock.[12] People were invited to participate in the ongoing progress of the year. It was their business, and a matter of life or death.

The 1857 event included backsword and wrestling contests; plus foot and hurdle races, and a special race down the Manger 'in pursuit of the cheese'.[13] This resembled the precipitous cheese race down Cooper's Hill, in Gloucestershire, which still occurs in late May, and combines a solar image with a dairy product.

At Uffington, the beer tent was of sail cloth, supported on 18-feet (5.5 metre)-high fir poles, erected in the nearby hill fort. It contained

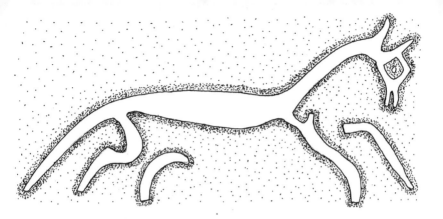

The horse deity Epona, widely recognised in the Uffington, Berkshire, hillside image, carved in 1000 B.C. and maintained ever since.

a double row of 18-gallon casks of ale, smaller casks of spirits, and a table laid with great joints of beef and pork, with baskets full of loaves, potatoes and lettuce. The side stalls, neatly arranged in an avenue down the Manger, were 'flower-bedecked booths', selling nuts, apples, gingerbread, toys, ribbons, knives, and other trifles, around which stalls numerous 'musicians and acrobats' performed.[14]

At the 1780 event, *The Oxford Journal* advertised, for 2/6, 'a ride down the Manger on a horse's jawbones'. Dead or alive, the white horse was central to the drama, and to keep her image as visible as possible, her limbs, sharply defined by trenches filled with chalk rubble, were (and still are) regularly scoured clean by local villagers.

Consequently the restored horse could watch the wheel, probably aflame, as it trundled the entire 600-metre length of the Hrungrpyt, to finally roll across the lower Icknield Way (an ancient track, linking the Neolithic flint mines of Norfolk, via Avebury, to the south coast); after that to drop into the steep wooded combe, cut by the Ock's spring waters at the foot of the Manger. In doing so, the solar wheel would link parched earth to the fast flowing waters, emerging in white jets, that served the entire Vale, and the string of villages, (namely Longcot, Fernham, Baulking, Stanton in the Vale, Goosey, Charney Basset, Garford, and Marcham), now sited along its banks. The Ordnance Survey marks three springs in this Manger foot group. After the wet winter of 2012, no fewer than nine were clearly visible.

The Ock is named after the Celtic salmon,[15] Old Irish *eo*, and Welsh *eog*, here adopted into the English, as *ocenne wyllas*, 'Salmon Springs', as recorded in medieval documents. Not until the 19th Century was the Ock's official source moved four miles north, because the infant Ock had by then been sunk underground, into culverts, for a few miles north of Woolstone village. The original Ock runs 24 miles, from just below the Manger, to join the Thames at Abingdon, where salmon swamed until 19th Century pollution drove them away.

Prior to that, given the salmon's habit of seeking out headwaters in which to mate, the Ock's source may well have been the site of some egg-laying rudds, which eventually offered a new start for that fish's ocean-wide travels. In Irish myth, the salmon is regarded as the sacred epitome of wisdom and health.[16] After one salmon offered himself to a queen, she ate the fish, and in doing so conceived and bore a healthy child named Tuan, (Irish 'wholeness'), the progenitor of a new race.

As if to reflect this legacy, the Saxon word *eoc* means 'safety, help, succour, increase'. At Uffington, fish, horse and humanity shared a sacred world.

Folklore of the Horse

Further Celtic influence is traced in connecting the white mare on the hill, (as she is generally called) to an embodiment of the horse goddess, the mare, Epona,[17] (whose name, known across Europe, combines proto-Celtic *ekwos*, 'horse', plus the -*on* syllable denoting deity.) In Wales, her counterpart, the horse-goddess Rhiannon, marries the prince of Dyfed in west Wales. She then gives birth to both a boy, and a male foal, delivered annually on May Eve, the start of the Celtic summer half of the year.

But where were the Uffington horse's foals delivered? No-one has mentioned her offspring, since Ralf de Deceto's version of The Wonders of Britain, c. 1100, spoke of the White Horse of Uffington and her foal.[18] She also appears, flanked by two foals, in a bronze sculpture of British Iron Age date (now in the British Museum). Perhaps the foal(s) were engendered beneath what is now called Dragon's Hill, a dramatic, truncated natural cone, rising close to, and directly beneath her hind quarters.

So perhaps we should look again at that curious 197 metre high, natural tump, with its artificially flattened top, which could be read as her conveniently placed 'pregnant womb', modified for ritual convenience. Subsequently maligned, following the Christian take-over, her womb-hillock is now said to be where St George killed the dragon – a reliable sign of pre-Christian sanctity. Here the nation's Horse, converted into a warrior's mount, is asked to help kill a harmful monster, dispatched on the 'birth mound' then renamed Dragon's Hill, by the newly adopted patron saint of Britain. To drum home the message, a large figure of that saint appears as a statue on the porch of Uffington parish church.

By contrast, an Uffington Charter of A.D. 931 names the Hill Eceles Beorh,[19] combining O.E. *ece-lic*, 'eternal, perpetual, everlasting' and *beorgan*, 'to save, protect shelter, preserve'. This older title reclaims the hill's positive value and connects it to the equine mother image, as an abiding protector of new life, lying so close to the eternal womb-hill. And she was probably so positioned deliberately to take advantage of the topographic opportunities, jointly offered by Manger, springs and knoll. They represented an unusually fortuitous cluster of features; a telling landscape, that eventually inspired the White Horse response.

To find a parallel for this suggestion, one need look no further than Silbury Hill, 15 miles down the Ridgeway.[20] It provided a Neolithic prototype for Uffington's complementary Bronze Age replay of a related theme. Here, as at Silbury, a Harvest birth-event is the core of the drama. The foal remains incarcerated inside the 'Dragon Hill' until the *Gwyl Aust* festival, held on the August Quarterday. Hughes describes Dragon Hill as 'a curious round, self-confident fellow, thrown forward from the range and unlike anything near him'. It may also be more than chance that the farm lying immediately north of 'Dragon Hill' is aptly named Sower Hill farm, a surviving positive association with harvest.

In regarding the 'Dragon Hill' as a former 'womb', set as an intermediary between chalk lands and the Vale, the mare may have been presenting her forthcoming offspring to that Vale, to help promote agriculture in its every field. Likewise her Iron Age figurine clearly shows corn in her foals' mouths.[21] She also carries agricultural implements, and a cornucopia, to match the Welsh Caseg Fedi, 'Harvest Mare'. Epona's scope likewise combined the equine with a wide range of fertility functions.

Although the midnight ceremonies linking the Uffington foal to the mound are no longer re-enacted, a medieval parallel survives in an account of Rhiannon, Epona's Welsh counterpart. She mysteriously delivered a new born foal and a human baby to the doorstep of the king of Gwent, Wales' eastern province, on May Eve.[22]

Meanwhile, at Uffington, the Saxons rededicated an adjacent Neolithic long barrow to Wayland, their mythic smith, to equip their horse Goddess with the shoes and metal implements needed for her continuing work.

Every August, under her eye and influence, the crops in her White Horse Vale ripen, turn to gold, and stand ready to be reaped, just as they did in 1000 B.C. In return for this annual favour, peasants from that Vale regularly and willingly scoured the White Horse image clean. To them she was more than an old sign, but rather a multivalent symbol, the working image of their immediate and long-term prosperity.

In Wales, as in Ireland, every winter, the white horse emerged from hibernation as The Grey Mare. In her skeletal form, the Welsh called her the Mari Llwyd. She was paraded from door to door at midwinter; her head made of a horse's skull, her body that of a living man, hidden under a white sheet. His companions sang, begged entry into houses, and were rewarded with beer. So sustained, she was ready to return as the foal and child, delivered annually to a king on May Eve, by queen Rhiannon.

In west Wales, the Caseg Fedi, 'Harvest Mare', turned into the Hag, who was seen in the last sheaf to be cut in any year. This sheaf was plaited into three strands, and the harvesters then took turns to throw their sickles at her. The man who succeeded in cutting her down would then announce: 'Early in the morning, I got on her track, late in the evening I followed her, I have her, I have her!' (In a claim suggestive of mating). To which the other reapers asked: 'What did you have?' and the reply was 'A hag, a hag, a hag!' So the mare became her wintry equivalent, valued as the fearsome essence of continuity. The man's challenge was then to carry the hag, hidden under his clothes, into a house, and in a dry condition, passing women determined to drench this effigy before it crossed their threshold. If the man succeeded, he was given free beer or a shilling.

When the white horse appeared in English Mummers' plays, onlookers often tried to steal a hair from its tail, for a year's 'good luck'. In Lincolnshire the man inside the 'horse' wore a grain sieve around his waist, supported on shoulder straps.[23]

And in Padstow, Cornwall, the central feature of the May Day ritual is the mating of the Hobby Horse with a 'teaser', as the horse is sung through the streets.

So, in this dispersed manner, the Uffington horse completes its annual cycle. Straddling the years, the 'sun horse' earns appreciation and gives all round benefits in return.

UISNEACH HORSES REVISITED: MARE AND STALLION EFFIGIES UNDER THE HILL OF UISNEACH

'For O, For O, the hobby horse is forgot'.
Hamlet, Act III, Scene ii

In pre-Christian Ireland the country's four provinces were believed to unite within a fifth province, termed Mide, which was synonymous with the Hill of Uisneach in Co. Westmeath, Ireland. Mide's 'four-into one' inter-provincial fusion also brought together the cardinal directions, based upon sunrise (East = Leinster), noon (South = Munster), sunset (West = Connacht), and midnight (North = Ulster). These provincial 'sun-land' mergers were also aligned with the human body; thus East = face, West = back, South = right hand, North = left hand, with all the parts united in the Irish word *mide*, meaning 'neck'.

Hence Mide is where the supernatural powers bring everything into a dynamic unity. At Uisneach this was symbolised by a great

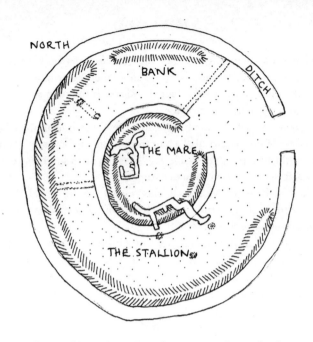

NORTH

BANK

DITCH

THE MARE

THE STALLION

*The Stallion and Mare horse-shaped souterrains, dug under the
'Horse Temple' at Uisneach, Co. Westmeath, Ireland.*

limestone boulder, named Aill na Mirrean, 'The Stone of Divisions';
several prehistoric earthworks, erected on the same Hill; and thirdly
an all-Ireland Beltaine (May Day) festival assembly – an Irish
Oenach[1]. As the hub of Ireland's five-fold structure, Mide was 'a
major ritual centre during the pagan Iron Age' and 'throughout the
Middle Ages'.[2]

After their successful excavation of the Neolithic temple on
Uisneach's summit, in 1929 the archaeologists Macalister and Praeger
turned their attention to another of Mide's grass-covered monuments,
and termed it Site 17.[3] They concluded that it was built during the Irish
Iron Age, c. 500 B.C. – A.D. 500.[4]

Site 17 (above) stands on the northeast end of the Uisneach hill,
and it consists of two concentric, approximately circular, banks, raised
from matching external ditches, surrounding a natural knoll. A single
entrance through the concentric banks was by a causeway, aligned
64 degrees east of north. This causeway passed over uncut sections

184

of the two ditches. The 250-feet (76-metre) diameter Outer Bank is now heavily eroded and its ditch, clogged with silt, now shows only as a 'shallow depression in the ground, yet the vertical drop from bank crest to ditch base was originally not likely to have been less than 18 feet'.[5]

Although the Inner Ditch has also silted up, the 120-feet (37-metre) diameter Inner Bank is better preserved. It encloses a 20-feet-high hillock, the flat summit of which was partly paved. (The footings of former walls that now subdivide the central enclosure into six smaller unroofed spaces belong to a much later era and distract from the prehistoric design.)[6]

Underneath the central knoll of this prehistoric monument Macalister discovered and recorded two stone-lined and flagstone-roofed tunnel systems or souterrains; but he did not attempt to account for their peculiar and elaborate shapes.[7]

More than 2,000 souterrains have been discovered in Ireland, ranging in date 'from the Late Bronze Age, c. 1000 B.C., to Medieval times'.[8] In functional terms they are usually described as storage cellars or hiding-places. However, as the archaeologist Charles Thomas wrote in his survey of souterrains on both sides of the Irish Sea, 'we cannot blind ourselves to the potentially sacral character of any pit, shaft, or cave in the insular Iron Age'.[9] In my book *Mythic Ireland* I interpreted Site 17's souterrains as a pair of hollow monumental images, depicting pre-Christian Irish deities, in their familiar 'stallion' and 'mare' forms.[10] These 'horses' are deliberately positioned close together, as if about to 'mate'. The Stallion, with his rear leg extended, measures 85 feet (26 metres) from hoof to tip of head. The 'mare' is 53 feet (16 metres) long from head to a contracted back leg. Both effigies were designed with roofs high enough to allow human entry into the bodies of these 'equine deities'.

Here I follow up that 1992 suggestion by adding that Site 17 as a whole may have been shaped to stage mimed 'gallops' of the sun and moon's circuits, envisaged as 'divine horses', travelling around, above and below the ground, in their yearly circuits. O'Riordain points out that this monument, with its concentric banks and ditches, is of a type 'developed in the Bronze Age' and often misleadingly described as a 'ring fort', since most of them 'had no military significance'.[11] Here it seems more likely, especially in view of its Uisneach location, that Site 17 was planned and built into a recognised *sid*, a supernatural hill, already inhabited by *aes side*, 'supernatural beings', in equine shape.

Above ground, the rath or ring-shaped earthwork's concentric banks may well have evoked sun and moon, with an additional solar-lunar 'chariot wheel' connotation provided by the two radial walls that join Inner and Outer banks.[12] In Irish Prehistory, visual metaphors were habitually mixed.

Beneath the central hilltop enclosure of this Irish *sid*, horse deities, free to roam from below ground level to sky, were stabled in a pair of horse-shaped tunnel systems, ready to mate. Therein, accompanied by human 'horseplay', a sense of union was engendered with the homeland of Ireland, experienced as the deity Eriu, 'a goddess with many horse-related features', in Puhvel's words. These included an ability to gallop through seasonal changes and to offer herself as 'Mare of Sovereignty' in kingship rituals.[13]

In its entirety, Site 17 was involved with cyclical movement – diurnal, monthly and annual, endlessly repeated. The temple stood for the supernaturally granted heartbeat of the common good; hence Ir. *sid* means 'peace, goodwill, a truce'. With this in mind the monument's 'circular earthen banks' make an Irish rath; and that word carries extra meanings; 'divine grace, virtue, a gift, good luck, guarantor, fortune, prosperity'.[14]

In Irish folklore supernatural horses played an active role in the Beltaine festival, centred on the mid-point or 'quarter-day' between the March equinox and the June solstice. Beltaine celebrated the start of the summer half of the traditional year at Site 17.

A Union of Opposites

When Site 17 was built the universe would have been seen as a supernatural living being; hence Irish *sùil*, 'eye', also means 'sun'. At Uisneach, solar and human eyes combined to enjoy the sight of Ireland as the goddess Eriu, alias Eire.

Although Uisneach is only 602 feet (183 metres) high, in good weather a remarkable 360 degree panoramic view from her summit can reach across her Central Plain, to the mountains beyond, which (as some of their surviving place-names indicate) evoke her gigantic breasts, hips and knees.[15]

To balance this horizontal array, a vertical connection between sky, earth and underworld is made by the Site 17 souterrains. Given a prehistoric belief in the Earth-centred rotation of Sun and Moon, Ireland's solar and lunar 'horses' were seen to arise daily from the sea, prior to leaping in luminous arcs over the world, before dropping below the horizon's dusky 'underworld' again.

The 'fairy hill' at the core of Site 17 contains an architectural show of the underground phase of this orbiting drama. There the celestial nags, depicted in tunnelled form, gather strength prior to their next circuit. For these immortal steeds there was no winning post, but only an endless race that linked abyss to zenith – a feat that (if respectfully acknowledged) might bring prosperity to hunters, herders and farmers, year after year. 'By such sights the deities surprise us. In this manner they proclaim their unfaltering nearness,' as Heidegger remarked, in 1971.[16]

By galloping between extremes, the Irish horse-gods embodied the union of opposites: light buckled to dark, hot biting cold, and life wrapped up by death. Their journeys also brought a free exchange of form between every kind of creature 'under the sun', as described in the adventures of the Irish sun goddess Etain, whose name stems from Ir. *etan*, 'poetry'.[17]

Poetry often deals in metamorphosis between species, and volatile exchanges between the four elements. So, after a jealous rival transforms Etain into a pool of water, 'the heat of the fire and the air and the seething of the ground... turned [the pool] into a worm... and the worm became a purple fly, big as a man's head, the comeliest in the land.'[18]

As a huge fly Etain travels over Ireland, but she is unable to settle, except 'on rocks of the sea and the ocean waves', until eventually she is given a *grianan*, 'sunny place' or 'sun-bower', on land, with 'bright windows for passing in and out'. But then a blast of wind drives her into the sky over Ireland for seven years, till weak and faint, she falls into a drink-filled gold beaker and is accidentally swallowed by a woman who thereby becomes pregnant and gives birth to another version of Etain.

In subsequent 'deaths and resurrections' Etain becomes a swan, and a beautiful young woman. Her surname, Echraide, 'Speedy Horse', links all these changes to a sun-horse pedigree described in the horse-shaped souterrains at Site 17. Within those hollow stone effigies, aspects of Nature's endlessly transforming processes could be re-enacted in mythic terms.

Etain's husband is the horse-god Eochaid Airem. His *airem*, 'plough', derives from Ir. *ar uaimh*, 'ploughing caves': 'It was he who first dug the earth to make a cave in it'.[19] He had learnt the knack of yoking oxen for ploughing from Midir's *sid*, Bri Leith, Co. Longford. As a 'first farmer', Eochaid Airem lived inside Frewin Hill,[20] eight miles from Uisneach. A fabulous example of his *ar uaimh* might be seen in the Site 17 souterrain 'caves'.

To dig the huge 'furrows' of the surrounding near-circular banks and ditches likewise called for a superhuman effort. Macalister described the cutting of the 800-feet (244-metre)-long and five feet (1.5-metre)-deep outer ditch with the tools then available as 'a stupendous task. The ground is so hard that a strong man, armed with a modern pick, finds difficulty in making any impression upon it'. Yet (divinely inspired?) the prehistoric community completed the job, albeit with a slight deviation in the northwest, to avoid a particularly hard patch of ground; and in so doing they (almost) simulated the rotating track of their gods.[21]

Sun, Horse, and Sid

Sunk below ground each night, Ireland's solar deities warmed the soil into potential fruitfulness for farmer and herdsman. This sky-earth collaboration is found in the goddess Grian, for Ir. *grian*, 'sun', also means 'earth, land, gravel, sand, sea or river bottom, basis, and foundation.'[22] When submerged, Grian lived inside a 'fairy hill' (Ir. *sid*) named after her – Cnoc Grene, in Co. Limerick. Nearby, her 'sister' Aine, sometimes known as Lair Derg, 'Red Mare', dwelt under Cnoc Aine, from where she emerged at the midsummer solstice, in time to lead a fire festival around her hilltop.[23]

Another Munster deity, the horse god Donn, lived inside Cnoc Firinne, the 'Hill of Truth'. From there people often saw him gallop upwards into the clouds, truthfully emulating the dawn sun rise, to speed for thousands of miles, envisaged in the shape of Eurasia's fastest animal, across the sky's vault.[24]

Across Ireland, from Dublin Bay to Galway Bay, a solar stallion, often named Eochaid in early manuscripts, repeatedly traced and embodied the sun's east-west course, from sunrise to sunset, at the March and September equinoxes. In the east, this steed appeared from the sea as the horse-eared Labraid Loingsech (literally 'Speaker Sea Rover'), founding father of Leinster. Twelve hours later his sunset counterpart, named Labraid Lorc, (Ir. *lorg*, 'course, track'), sinks due west into the *sid* named White Horse Hill, (Cnoc an Ghervain Bhan), near Clarinbridge, Co. Galway.[25]

Eriu's Horse's Hide
Wild horses have been absent from Ireland since the last Ice Age; but domesticated horses were brought here from Britain during late Neolithic times, c. 2300 B.C. Numerous finds of horse bits and other harness items show that horses subsequently played an important part in Irish culture during the Bronze and Iron Ages, with their supernatural roles echoed in recent folklore.

The idea of Eriu as the equine embodiment of a sun-land fusion features in Codal's medieval Irish legend, which says that in her infancy Eriu was clothed in a 'horse's hide' (Ir. *codal*), given to her by a horse named 'Codal the round-breasted': 'Tis he that was fosterer to Eriu... and on yon peak [Benn Codail] he used to feed his fostering. And every vigour that he bestowed upon her used to raise the earth under them.' So Eire, as the ever-growing Mount Codail, ascended steadily sunwards.[26] Alarmed, she begged Codal to stop feeding her and complained: 'I am heaved up so high that the sun scorches me!' He agreed, and as a result she was able to keep her distance from the sun, giving Ireland a restrained topography and a temperate climate.

The gold ore found in Eriu's Wicklow and Mourne Mountains was mined and purified from c. 2000 B.C. onwards, and turned into 'sun-wheel' discs, engraved with an equal-armed cross, to express the sun's movement. A pair of these was found at Coggalbeg, Co. Roscommon,

with a crescent moon-shaped 'lunula' collar, of which more than 80
Late Neolithic Irish examples are known.

Eriu at Uisneach

The people of pre-Christian Ireland assumed that everything in the
cosmos was fundamentally unified and able to metamorphose into
a range of different forms. Thus a deity could appear as a horse,
an island, the sun or moon, or in human shape.[27] So, while Site 17
contains divine horse images, the shape of the Neolithic temple, set
on the summit of the Hill of Uisneach, displays the goddess Eriu's
outline in superhuman shape. And it was there, says Lebor Gabala
Erin, that Eriu negotiated with the Sons of Mil, mythic warriors
who had swept into Ireland from continental Europe, introducing
iron swords and horse-drawn chariots, c. 500-300 B.C.[28]

Scholars now think that these 'Celtic' newcomers were few in
number, rather than leaders of a mass migration; yet, according to
legend, Eriu was unable to drive them out, and at Uisneach she agreed
to let them take over her entire country, providing they kept its Eriu
name. The Sons of Mil accepted her proposal and perhaps celebrated
the deal inside Site 17's subterranean images of the horse deities with
whom they identified.

Yet in doing so they inadvertently gave additional honour to the
incumbent An Dagda, 'The Good God', alias Eochaidh Ollathair,
'Horse All-father'. His Dagda title is derived from the proto-Indo-
European phrase Dhago-deiwas, 'shining divinity', cognate with
English 'day', and punned with Irish *dago-s*, meaning 'good', used in the
sense of 'extremely skilled', or 'good at everything' including building
raths, controlling the weather, and farming. The Dagda was installed
long before the Iron Age, among the Tuatha de Danaan, a pantheon of
divinities including Aine, Graine, and Eriu. The pantheon was named
after the goddess Danu, alias Anu, who shows her mountainous Paps
of Anu in Co. Kerry. Probably established in Neolithic times, the
Tuatha de Danaan's power continued to be acknowledged deep into
the Christian era.[29]

Irish culture throws little away, and Site 17 probably represents
an intermingling of 'new' and 'old' divine powers, with the Tuatha de

Danaan acting as supernatural guarantors of Milesian legitimacy. In the same way, Eriu's temple on the top of Uisneach Hill, 'the scene of great feasts and fires', was, in Macalister's view, 'very likely inherited by them' from the native population.[30]

Such seasonal fire rites continued into modern times. For example, in King's County 19th Century people jumped over a midsummer bonfire, hoping to absorb the power of that 'sun-in-miniature', as they momentarily 'became' the solar horse, during its highest annual leap. Then the participants greeted the arrival of a prancing Hobby Horse with cries of 'The White Horse!' before running off in all directions, laughing and screaming, as it pursued them, to challenge the leapers' presumption'.[31]

Long after Christianity came to Ireland, sun worship continued to involve supernatural horses, bonfires, and prehistoric monuments, including, perhaps, some of Site 17's features. Thus rath is related to Ir. *roth*, 'wheel-shaped disc, or sphere'. The monument's banks offered tracks for a horse, circuiting in imitation of the apparent movements of sun and moon. The radial walls (added later) join the Outer Bank's 'rim', like the spokes of a celestial deity's 'chariot wheel', as displayed with a solar horse on Iron Age Celtic coins.

The archaeologists found that a 'red virgin earth', sunset-fiery in hue, characterised the Outer Area (stretching from Site 17's Outer Bank to Inner Ditch). This colour matches An Dagda's 'eye' at sunset, that gave him the extra title of Dagdai Dierg, 'Red one of great knowledge'. In addition, Macalister found 'traces of large fires', set in and around a 'trilithon megalithic doorway'[32] giving access downwards into Chamber A of what he called the 'Lesser Souterrain'.

The Stallion Souterrain

In this account the chunky stone doorway is equated to the Stallion's rear hoof. The 'hoof-as-auspicious-doorway' formed the flame-surrounded threshold through which the solar stallion landed on Earth, leaving his print on the ground, as he propelled himself into the dusky abyss. (As the visual symbol of this liminal event, horses' hooves were venerated in Ireland. 'When a horse is dead, they hang up his feet and legges in the house, yea their very houfes [hooves] are esteemed as an hallowed and sacred relique', wrote William Camden, in 1607.)[33]

At Site 17 the narrow 2' x 2' (61 cm x 61 cm) 'rear hoof', capped by a stone lintel, forms the entry into the supernatural Stallion. It leads to a passage within his hind leg and beyond that, into the entire horse-shaped souterrain. This hollow effigy is aligned sun-wise (Ir. *dessel*), on a perpetual 'journey' from south to north, and from daylight into the dark subsoil.

In order to enhance the dynamic quality of the Stallion's arrival and plunge into the earth, the much shorter original 'rear leg', defined by Chamber C, was extended by the construction of Chambers A and B. They added more than 33 feet (10 metres) to the back leg's span.[34] This change, perhaps ordered by 'The Sons of Mil', involved breaking through a section of the previously dug Inner Bank, and re-filling a section of the 8-feet (2.4-metre) wide Inner Ditch.[35] These alterations enabled a newly supercharged 'Stallion', with one full-stretch leap, to unite the sunset-red Outer Area, from which he now sprang, to the Moon's realm, rimmed and defined by the Inner Bank's 'lunar circuit' and equated in Irish folklore with the Lair Bhan, 'the White Mare' of Sovereignty. The extra leg length also showed Eochu Ollathair as Dagda Dein, 'Swift Dagda', a frequently used epithet.

Chambers A and B display in their plans the curves typical of La Tene late Iron Age design, found on pots and metalwork of the period. The tunnels are floored with tramped clay and both souterrains have walls of un-mortared local limestone slabs, of a silvery grey hue. The Stallion's hind leg walls stood three to four feet (0.9 to 1.2 metres)-high in 1930, though by that date all the roof slabs that once covered these two 'limbs' had been robbed. The 'lower leg' is 17-feet (5.1 metres)-long and five feet (1.5 metres) across. A sill of three stones defines the 'knee' and start of the steed's 16-feet (4.9-metres) long 'upper leg'.[36]

Chamber C describes the Stallion's 16-feet-long original hind leg, which retains its own substantial megalithic 'hoof-doorway', less than 2-feet-square, and set entirely beneath the Inner Area's surface. Through this aperture, and by crossing a 6-inch (15 cm)-tall stone ledge (simulating a 'fetlock'?), the pilgrim may walk, ducking to avoid some sharp points in the overhead roofing slabs, towards the Stallion's back.[37]

Dullahan Nightmare

To be inside the solar horse-god was no doubt an awesome privilege – 'a dream come true'. Eurasian folklore continues to maintain that it is lucky to dream of horses, while from China to Ireland, in Antiquity, the horse was credited with magical powers and the ability to carry people to the 'Other World'. Here at Uisneach, entry into the Stallion souterrain was probably reserved for special moments in that deity's annual routine, when the pilgrims (or those acting for them) might best merge with the steed's tunnelled body and persona.

Yet this was a perilous, fear-filled undertaking, for however pale the limestone material used to make these underground passages was, to those who fumbled along the hind leg walls and groped towards the 'sun horse' body cavity, the hollow steed was dark as moonless midnight. Had visitors been tricked into a premature grave, victims of the dreaded Dullahan?

In Irish folklore The Dullahan is a dark, headless, fairy horseman, who appears most often at Samain. He wields a whip made of a human spinal column and is mounted on a jet-black, fire-breathing steed, with flaming hooves. When this horse stops outside a house, its rider's decapitated head calls out the name of a sick person lying within, who gives up his soul and dies.[38]

The only defence against The Dullahan's summons is to wave something gold, no matter how small, at the horse; whereupon horse and rider will vanish with a roar, leaving a fiery trail in the air. As the shadowy counterpart of the solar steed, The Dullahan serves as a reminder that some form of his own mouldy, wide-grinning, head, carried under one arm, addresses and ends every mortal life and all imaginings, including the very idea of solar horses. As if to acknowledge that any concept of Totality is based on a combination of opposites, The Dagda, as the supreme solar stallion, was sometimes called the 'Dark One', Dagda Donn of the 'dark cloak', (Ir. *lachna donn*). His sun-horse brightness created its own shadow.

In modern astronomical understanding, another version of 'The Dullahan' cries 'Black Dwarf!' at our dying sun, before tipping its cinders into darkness. Today, as in the Iron Age, we are mounted within the body of galloping transience. Those who entered the underground Stallion knew and experienced that condition.

Cauldron of Plenty

Chamber C of the Lesser Souterrain ends in a right angle bend at the Stallion's hip, where a lintel (partly dislodged by 1930), leaves barely enough room to squeeze into Chamber D, the animal's flagstone-roofed 29-feet-long 'croup, loin and back'.

What the excavator termed a 'small cupboard-recess', Chamber E, projects from the underside of the Stallion's loin, as if to simulate his scrotum and also perhaps the never-empty 'Cauldron of Plenty', attributed to The Dagda in Irish myth.[39] Demonstrating its flexibility of scale, the same magic vessel was often equated with the firmament's star-studded vault.

Sited directly above Chamber E a deep ash pit was discovered, dug into the surface of the courtyard, together with a paved hearth, and near a 10-feet (3 metre)-long line of big stones, possibly representing on ground-level the extended phallus of The Dagda, in his 'Horse Great-Father' persona, when aroused by the nearby Mare souterrain.[40] On and below ground the Stallion's sun-derived virility was celebrated and encouraged.

The interplay between celestial, surface and submerged realms, as featured around Chamber E, finds a verbal parallel in descriptions of The Dagda's exploits, given in the *Second Battle of Mag Tuired* manuscript.[41] There, although he retains his horsehide shoes (with the hair explicitly worn on the outside), he adopts a superhuman shape; with his 'uncovered penis' doubling as his 'mighty club, with which he digs the boundary ditches between provinces'.

However, when he meets and lusts after a young woman, his huge belly, swollen with the fruits of the previous year's harvest, renders him impotent, so the maiden disdainfully hurls him into the ground, up to his rump. There he releases the contents of his stomach, climbs out of the hole, sets the girl on his back, and in so doing displays and drops his 'stone testicles' (the seeds of a future harvest?), before the couple eventually make love.[42] These comic exploits match the rain-drenched uncertainties of the Irish farmers' year, which the divine pair underpin and initiate. Having been mocked and spurred on by the young goddess, The Dagda can now (with first-hand subsoil experience), keep the agricultural cycle turning.

As regards The Dagda's horse-to-human transformations, modern science has recently established that we share at least 62 percent of our genetic code with the horse. So the wall erected between humanity and other genera is found to be permeable, as in pre-Christian times, when the Site 17 horse effigies freely expressed Nature's underlying unity.

To demonstrate this transfiguration, people might have entered the Stallion's souterrain 'gut' in a mood of give and take, willing to serve as symbolic 'grass' or 'hay'; fodder for the gigantic herbivore, who for them integrated everything under the sun, to Ireland's advantage. The Irish word *capell*, 'horse', also means 'foundation'. Upon this horse the Irish loaded their hopes. With such a range to cover, can one doubt that larger-than-life-sized horses were imagined in Ireland – sometimes, as at Site 17, more than 80 feet long, where horse-deity rites could be enacted?

From the Stallion's shoulder his foreleg (Chamber F) is more than 14 feet in length, 'shapely, and built of fairly large stones'. The foreleg contracts at the south end into a megalith-defined 'hoof' which (like the rear hoof), is surrounded by a bed of ashes, from celebratory prehistoric fires, signifying that, from sunrise to sunset, the solar-horse planted his hot hoof prints into the earth.[43] The steed's 'neck' and 'head' are shown combined in the 14-feet-long flag-roofed Chamber G. In plan this begins with a 'round-breasted' bulge and then narrows at the 'neck', through a stone doorway, before gradually expanding as it approaches his upraised 'head' and broad 'muzzle'. His nose comes within 14 feet (4.3 metres) of the 'Mare's' hindquarters, just as a stallion normally sniffs at a mare's tail, dock, and urine, prior to an attempt at mating.

The 'Mare' Souterrain

Macalister's 'Greater Souterrain' is here recognised as The Mare. He labelled the chambers A to K, reading from her 'head' to her 'hind leg'. This hind leg is deeply sunk, matching the winter part of her yearly journey. Between the 5 feet (1.5 metre) high 'very rough' stone walls of Chamber K, her rear 'hoof' and 'fetlock' were anciently defined by two pits, dug into its floor. Visitors were saved from asphyxiation by a 13 feet-long air-vent shaft.[44] Flexed at the 'hock', the Mare's two hind leg chambers have a combined length of 26 feet (8 metres) and display a stationary pose, as if she is waiting to be covered by the Stallion.

A pair of massive 4-feet-high stone uprights offer an entry below the Mare's 'dock', or fleshy stump of her tail, after which the floor rises by three feet, involving a crawl over a large stone block, into her croup and hip, defined by Chamber G. Here Macalister noticed 'a small [stone] cupboard', set into the chamber's east wall, 'possibly for holding a lamp to illuminate this dangerous part of the cave'.[45]

Such a light could also stand for the sun Stallion's entry into the Mare, making this passage of considerable interest to those seeking to identify with their act of union, as shown by sections of the side walls of Chamber G that had been 'carefully pointed with grey puddled clay, neatly smoothed and pressed in, with imprints of fingers and a whole hand of a child or woman'.[46]

In addition, the archaeologist noted 'extensive polishing on the stones of this cave', with the walls 'polished smooth' in places. Puzzled, he wrote: 'It is hard to see why there should have been such intensive traffic through a passage which seems to lead nowhere'.[47] He wondered if penned sheep had been responsible for the polishing effect, but the farmers he consulted said that flocks would never be housed in such a narrow passage for fear of their suffocation.

Once the souterrain is recognised as a monumental sacred Mare, the necessity felt by women and girls 'to keep in touch' with the creature's vital organs in order to benefit from such contact, in promoting their own fecundity, provides a plausible explanation for the polished walls 'enigma',[48] especially during an era when the divine mare could also assume human form, as was the case with Rhiannon, the Welsh horse goddess who sat beside a mounting block, and offered people a ride on her back.[49]

Far from 'leading nowhere', those who arrived in Chamber G occupied the Mare's rump, where the Stallion hoped to inseminate his 'on heat' partner. Just as the stone effigy of a much-revered saint turns shiny in patches from the outstretched fingers of countless supplicants, so here, the repeated human wish to derive help from physical intimacy with the great horse-goddess (regarded as mother of all living things), had smoothed the most 'private parts' of her rough stone walls.

The polishing was most pronounced close to the chamber floor, which suggests that the pilgrims may have attempted to mime equine mating, perhaps followed by an 'incubation slumber', entailing a dream-union with the horse deities, to enhance their own fertility. On hands and knees, they may also have imitated the posture of a crawling infant. Speculation about the uses made of Chamber G ought to go beyond modern functionalism's restricted causality, in order to re-enter the now unfamiliar realm of sympathetic magic, in which this effigy stands.

Roofed by huge slabs, the Mare's 'back' gradually rises in height from three to five feet, and is joined at right angles by Chamber F. The doorjamb at this critical junction is formed by 'a large rounded block of limestone containing much white calcite, and very highly polished over a square foot'.[50] It had presumably been selected for its whiteness, to epitomise the Lair Bhan, the sacred 'White Mare' (so familiar to Ireland's folklorists), displayed here in the midst of her cavernous generative organs.

The Mare's Uterus
Chamber F is interpreted here as the White Mare's uterus. It was constructed of smaller stones than used elsewhere. Positioned over its south end, Macalister found a vertical shaft, 5 feet (1.5 metres) deep and 3 feet (91 cm) across. (This shaft opened onto the surface of the Court above, close to where a cluster of seven substantial 'Maypole' post-holes had been sunk into the courtyard's surface.)[51]

Three stones projecting from the shaft's sides 'facilitated descent and ascent'. This shaft had been deliberately positioned over a natural vertical fissure in chamber F's floor, that Macalister excavated to a depth of nearly three feet without reaching its bottom, but was reluctant to delve further, for fear of damaging Chamber F.[52]

This perpendicular axis hinted at the Mare's scope. From her Chamber F uterus, she plumbed the abyss, over which her souterrain body lay. This line rose through the shaft to the surface of the *sid*, and from there to the firmament's zenith. Rather than an abstract 'concept', this vertical 'passage' was designed for use in delivering The Dagda's offspring, both equine and human.

Whoever or whatever emerged from Chamber F might also serve to re-enact the birth and successful delivery of a future harvest. A straw effigy, called Caseg Fedi, 'Harvest Mare', epitomised crops of wheat, barley and oats throughout Irish-settled areas of West Wales.[53] Plaited from the last sheaf of the year's crop, this 'mare' figure was stored in a barn through the following winter, before she was ploughed underground again in early Spring, as the initial 'sacred seed'. Thus on both sides of the Irish Sea the 'harvest mare' sank and rose, as a staple vegetable aspect of the Lair Bhan's wide-ranging cycle of achievements.

From a sub-soil presentation in her Mare-shaped Site 17 effigy, the Irish version of Caseg Fedi delivered the nutritious food of primal one-ness - an 'Earth-Sky-Divinities-Mortals' unity – onto every family table.[54] In order to do so, she incorporated solar and lunar power. Similarly, moonlight shines from the Welsh goddess Arianrhod, 'of famous beauty, surpassing the hue of sunshine'. Her name means 'Silver Wheel',[55] and her islet-home lies between Irish Leinster and the Welsh Lleyn peninsula, named by Irish settlers there.

Lair Bhan in Action

Ireland's prehistoric monuments cry out (sometimes in vain) for reconnection with that country's exceptionally rich folklore. So from her Site 17 effigy we may trace the White Mare's progress through her yearly exercise routine. She was a principal figure at the Midsummer bonfire rites. Seven weeks later, during the Lughnasa, First Fruits festival, her white-draped equine image trotted through Lammas Fairs on August 1st.

Three months later, at Samain, in Co. Kilkenny, a procession marking the start of the traditional New Year was led by a figure enveloped in a white robe or sheet called the Lair Bhan, who acted as a sort of president of the ceremonies. Also at Samain, The Dagda is said to have mated at Corran, Co. Sligo with his mare-shaped wife, Morrigan. Consequently, any wish that a person made on seeing a white horse at Samain was likely to be fulfilled.[56]

Just after the winter solstice, on St Stephen's Day (December 26th), the Lair Bhan reappeared as a 'hobbyhorse' that paraded through town and countryside. A boy hidden inside a wooden, cloth-covered frame operated her snapping jaws.[57] Horses were also involved in Ireland's February 1st quarter-day, when Brig, the prehistoric Irish 'fire' goddess (later cloaked as St Brigit), continued to drive her horse-drawn sky chariot from 'Heaven', in time for her annual 'rebirth' at the feast named in her honour.

The White Mare was the 'Sovereign Goddess' of Ireland in equine form – a role consummated and secured for her by The Dagda. His proximity to her at Site 17 confirms in architecture what Irish myth described in words as a coronation ritual. Humanity and mare become one in this 12th Century inauguration of an Irish king at Kenelcunill, Ulster, witnessed by Gerald of Wales:

'A white mare is brought forward into the middle of the assembly. He who is to be inaugurated... has bestial intercourse with her before all, professing himself to be a beast also. The mare is then killed immediately, cut up in pieces, and boiled in water. A bath is then prepared for the man by all his people, and all, he and they eat of the meat of the mare.' Then 'he quaffs of the drink of the broth in which he is bathed, not in any cup, nor using his hand, but by just dipping his mouth into it around him [as a horse does]. When this unrighteous rite has been carried out his kingship and dominion have been conferred'.[58]

'I eat you, you eat me,' is a theme that underlies many of the world's religions, including Christianity, says Joseph Campbell; and through her sacrifice the divine mare retained her eminence in Irish sovereignty lore, during which the king served as her 'Dagda stallion'.

In India, by contrasting inversion, the future Chief Queen (*mahisi*) was ritually impregnated by the sperm of an ejaculating stallion, who was smothered to death at the culmination of her *asvamedha* inauguration rite.

Coronation Mare

Versions of the Lair Bhan rite that Geoffrey saw in medieval Ulster were probably held throughout Eire during the Iron Age, and were likely to have involved the Site 17 effigies in the coronation of a high king, holding inter-provincial responsibilities. In return every chieftain who attended the Beltaine festival at Uisneach 'donated his horse and trappings', according to Keating.[59] Likewise those who entered the Site 17 souterrains' guts offered and in a sense, became the 'food' of appreciation, without which those carefully shaped 'horses', starved of recognition, would simply 'vanish'. And that is what eventually happened. Consequently no horses were 'recognised' there, during the 1929 excavation.

By contrast, after their coronation, many Irish kings, who feature either in Irish myth or as historically verified individuals, chose to incorporate the *eoch*, 'horse', element into their royal names. Mounted, and with a total of six working legs, capable of different speeds, like a car's gears, the monarch readily married an equine divinity, such as the Mare of Sovereignty, who dispensed some of her own cyclical immortality to strengthen the monarch's linear reign and invested him with the 'divine right' to rule her land. Her gift was firmly embodied, as working architecture, at Site 17, on Eriu's central hill.

The Mare's Womb

Resuming an unfinished exploration of the Mare souterrain's torso necessitates crawling through a narrow 3 feet (91 cm) high passage into Chamber E, her trapezoid-shaped capacious 'womb'. More than 10 feet (3 metres) long and 5 feet (1.5 metres) high (with only two of the roofing slabs left in position by 1930), it has 'awkwardly small doorways' at both ends, and a 'kidney-shaped hollow', 28 inches (70 cm) long, sunk into the floor.[60] This cavity was found neatly 'filled by a block of stone that lay within it'. The stone was probably

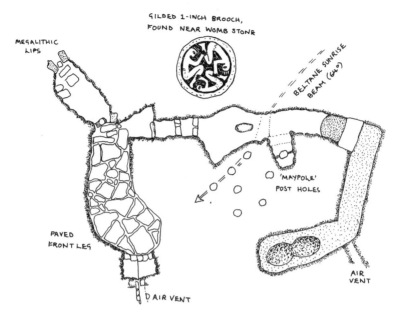

A detail of the 'Mare' souterrain, at Uisnech, Co. Westmeath.

intended as the Mare's 'unborn foal', sired by the 'Stallion' souterrain, and rediscovered by Macalister, safely enwrapped in its 'caul-hood' pit, where it had lain for more than two millennia. Here the immortal, repeatedly sacrificed Lair Bhan is shown in a perpetually pregnant state.

Perhaps for that reason in the horse-shaped Iron Age souterrain in South Uist, known as 'The Mare-House' (Gaelic *Tighe-Lair*), the Mare's 'neck' passage was found to be of 'incomprehensible narrowness', while her wide womb chamber had a vault 7 feet (2.1 metres) high.

This unborn 'colt' stone at Site 17 was matched, in superhuman terms, by the discovery in the same chamber of a bronze disc brooch, 1 inch (2.5 cm) in diameter. Macalister mistakenly identified this item as Anglo-Saxon. In fact it is engraved in an Iron Age Irish style, and displays three humanoid figures, curled up in foetal positions, endlessly rotating within their 'solar-lunar' circular rim, on which traces of gilding survive. All three figures wear caps and belts, yet clutch at one another's shins, in a possibly pre-natal state. Like the Welsh horse goddess Rhiannon, who gave birth to foals and boys, it seems that the Site 17 mare was credited with producing both.

Pre-Celtic and Celtic deities often come in threes and here the brooch figures may well represent the trio of gods named in Irish myth as Eriu's husbands. Her provinces merge at Uisneach, where she negotiated her country's future. One might therefore expect to find evidence of her spouses at Site 17, especially given the wintry 'underground' connotations conveyed by two of these husbands' names.

Mac Cuill, (Ir. 'Son of Holly') was married to Eriu-as-Banba, her earth delving, 'pig goddess' persona, (Ir. *banba*). In folklore holly often has male associations with midwinter fire. (At Brough in Celtic Cumbria the village strong man carried a holly tree around, with flaming torches tied to each branch, to welcome the next year.)

Mac Cecht, ('Son of Plough-share', Ir. *checht*,) was the husband (and husbandman) of Eriu-as-Fodla, (Ir. *fodlach*, 'a division, a part'). With the tool embedded in his name, he cut the first furrows into her earth, perhaps the boundary ditches of Eriu-Fodla's four provinces, that met and coalesced around Uisneach's Stone of Divisions.

The third god in this trio, Mac Greine, derives from Ir. *grian*, 'the sun', matching his wife's solar (Ir. *Ei*) 'regular traveller' element, contained in the Eire form of her name. With the Site 17 brooch, he married his sunny warmth to hers, inside the Mare's wedding Chamber E. That was where the gilded bronze disc brooch was discovered, after falling from the courtyard immediately above. Probably donated as an offering to the Mare, it was found with its pin and fastener torn off, signifying that it belonged to no one and everyone, within the eternal cycle of death and renewal. Along with his two brothers-in-law, here Mac Greine 'got down to earth' in the subterranean seedbed of the flying white horse.

Together this iconic brooch and mare effigy generated a trans-species progeny of various crops and animals. This multiple nativity was celebrated by ceremonies staged in and around the monument, to be echoed by mythic word and comparable events, including dance, simultaneously held at many other sacred sites, across Eire. The fruitfulness of Ireland was encouraged by these sympathetic gyrations.

Mare Emerging

Further progress towards the Mare's neck entailed a struggle along her back, over a stone sill that 'shows conspicuous signs of wear'. It gives access to Chamber D's narrow creep passage, less than 4 feet (1.2 metres) high, providing a 'rite of passage' for many generations.

After that, the Mare's 'withers' are defined by Chamber C, which connects with her capacious Chamber B 'front leg'. Neatly paved with flagstones, its breadth of up to 7 feet (2.1 metres) could accommodate those who had completed (or were about to start on) a journey through the tunnels. At its 'fetlock' end Chamber B finally contracts into a stone-lined 'hoof' alcove, from where a narrow air vent reaches the surface in the surrounding ditch.[61]

The mare's neck and head are combined into the 12-feet (3.6-metre)-long Chamber A. Its roofing stones (now robbed) doubled as paving for the overlying courtyard. Unlike her northward-aligned torso, Chamber A swings to the northeast, as if searching for something – perhaps her own manifestation as the divine sunrise horse who emerges annually at Beltaine dawn, thereby releasing the summer half of the year from her six months of wintry incarceration.

Accordingly the floor of her 'neck' slopes gradually upwards towards her 'head', from where stone steps ascend from the underworld and break to the surface at her open 'mouth'. This was formed by a lintel (now gone) and by two upright stone jambs. They are over two feet tall, and set 2 feet 3 inches (68 cm) apart – just wide enough for human pilgrim-priests or priestesses, perhaps dressed as 'horses', to pass from the souterrain onto the courtyard's surface.[62] In pointing northeast, her 'mouth' addresses the quarter of immanent summer sunrise, and announces the start of the annual Beltaine festival.

St Patrick Intervenes

When rediscovered in 1929 the 'Lair Bhan's mouth' had been 'stopped up with loose masonry'. Was it blocked as a danger to straying livestock, or as a consequence of St Patrick's 5th Century arrival on Uisneach?[63]

There is no doubt that he was bitterly opposed to the sun worship prevalent in 5th Century Ireland. In his autobiographical Confessio he writes: 'For this sun... will never reign, nor will its splendour last; what is more, those wretches who adore it will be miserably punished. Not so we, who believe in and worship the true sun... Christ... who will never perish.'[64]

Yet in his missionary struggle against all the Irish pagan sun deities, the saint himself is reported to have absorbed solar powers. For example, when his own charioteer lost his horses in the dark, Patrick's fingers flared like lanterns until the steeds were found.

Likewise he could turn icicles into flaming tinder, says the Vita Quarta manuscript.[65] According to Muirchu, the patron saint's 7th Century biographer, Patrick 'founded a cloister there' [at Uisneach] and invited two sons of Niall, Fiachu and Endae, to join him. But 'they refused and expelled him'. '"A curse," saith Patrick – "on the stones of Uisneach," added [his henchman] Sechnall. "Be it so," saith Patrick', thus passing responsibility for the damage inflicted on all Uisneach's stones to his own henchman. 'Nothing good is made of them from that time forward. Not even washing stones are made of them,'[66] is Muirchu's assessment. Muirchu added that the Uisneach Beltaine tradition was ended by St Patrick, for the saint had moved the festival to Tara, where he made it coincide with Christian Easter. Yet to this day, a 'holy well' and a 'bed' of loose stones on Uisneach Hill are named after St Patrick.

One attempt to accommodate Christianity into pagan Uisneach was made by King Diarmait in the mid-6th Century. Installed as High King of Ireland, he presided over Uisneach's Beltaine Convention, and persuaded St Ciarnan of Clonmacnoise to join him there. King and cleric sat side by side during that 14-day-long Dal Uisneach festival, and the king gave Ciarnan a grant of land.[67] Thereafter an uneasy co-existence between two incompatible outlooks was achieved throughout Ireland, for the rural inhabitants were unwilling to forgo their deep-rooted festivals, or the prosperity that the old deities were believed to deliver.

Consequently (according to Irish folklore) the patron saint incorporated supernatural horsepower into his own legendary steed named Gearrain Phadraig, 'Patrick's Nag'. This 'sorry-looking beast' retained a pre-Christian link with agricultural prosperity and was 'no doubt a god's attribute'; for it could 'carry all the corn in Ireland on its back, and in a single load! No burden was ever laid on its back that it was not able to carry.'[68]

Beneath this 'work-horse' disguise, the legacies of An Dagda, alias 'Horse All-Father', along with that of the Lair Bhan, epitome of 'Sovereignty', persisted, as they rode together through Irish culture towards modernity. And now that their battered souterrain effigies have been unblocked, they can both jump into the world again, to be welcomed anew as symbols of cyclical cohesion, without which Doomsday becomes a self-fulfilling prophecy. The Site 17 divine horses have accordingly been greeted with enthusiasm by the organisers of the Festival of the Fires – an annual Beltaine celebration, revived on Uisneach in 2009.[69]

Beltaine at Site 17

As defined by Chamber A, her stone portal, the Mare's open mouth suggests the Irish word *bel*, meaning 'cavity, orifice, lips, big-lipped'. Irish *bel* is also 'the mouth as the source of speech'. During prehistoric times, sounds and echoes heard in caves were often interpreted as the utterances of supernatural beings. So here, from her souterrain's equine throat, the White Mare's 'whinnying' (however simulated) could put the *bel* into Beltaine. By mythic word she simultaneously announced and produced the rebirth of the summer half sun. This occurred annually at dawn on the May quarter day, precisely halfway between March equinox and June solstice.

The Mare souterrain was as good as her word, for Ir. *bel* can also mean 'sparks from boiling heat' and, in this sense, an extra fiery demonstration of her upward-surging solar personality was cradled in a pit, dug into Site 17's courtyard surface only 10 feet (3 metres) from her souterrain's 'lips'. The pit, excavated in 1929, was 8 feet (2.4 metres) in diameter and 'contained a deep layer of ash, testifying to its long-continued use as a place of great fire',[70] with the surrounding courtyard surface also found to be 'black with fires'.[71]

There, on Beltaine eve, the divine Mare and Stallion metamorphosed into a bonfire, set below ground to symbolise the pre-dawn sun's 'flaming eyes', (Ir. *Lasairshuili*), from which her extended hot 'tongue' flicked into the night sky, to 'speak' of the first summer sun's imminent arrival. Thus from the horse's mouth, the souterrain-mare, having coupled with her stallion, could combine Ir. *bel* with Ir. *teine*, 'fire', to produce the festival named Beltaine. The solar-lunar Lair Bhan was often regarded as the source of all Eire's *teine*, or *taine*, 'fire'. At the Beltaine festival she gives her 'word', Ir. *bel*, that she would deliver *taine*, 'fire', from the six-month grip of winter.

Beltaine eve ceremonial fires were credited with supernatural protective properties; hence the Irish custom of driving cattle between two fires on that night, to protect them from disease during the coming 12 months – a practice that persisted far beyond the Iron Age.[72]

Bellowing cattle, stampeding between fires, fill the word Beltaine with extra physical action. This type of sacred word-thing-deed fusion is typical of myth based events, in which a blending of spoken liturgy, poetry, dance, song, costumed hybrids, painted and sculpted effigies takes place at a revered (super)-natural feature, sometimes enhanced by a 'temple' structure, as at Site 17.

There humanity tried to understand and to ask favours of the gods, in a process of mutual 'give and take'. Consequently Ir. Beltaine can also mean 'the food and drink sacrificed upon the May-eve fire'.

Bel-Beil-Belenos, or Belt-Aine?

To Keating, Ireland's 17th Century historian, the word *bel* was embodied in a god who he named Beil, credited with the power to protect cattle from disease when they, as tradition demanded, were driven between two fires at Beltaine. Keating's Beil was his version of the continental Celtic god, Belenus, alias Belenos, known from 50 inscriptions there, and sometimes equated with Apollo. A gold coin of the 1st Century British king Cunobelinus, meaning 'Hound of Belinus', shows a solar horse leaping over an image of the sun. Yet in Ireland, apart from Keating's reference, there is no record of Bel or Beil as a god.[73]

Horse and Beltaine Bonfire

That the solar White Mare physically blended with her May-eve bonfire is clear from 19th Century accounts from the Irish midlands where, 'a wooden frame, some 8 feet long, with a horse's head fixed to one end, and a large white sheet thrown over it, [was] ... greeted with loud shouts of "The White Horse"... With a bold leap, it was safely carried by the skill of its bearer several times through the fire. When the fire burned low, every one of the peasantry passed through it, [after which] children were thrown across the sparkling embers'.[74]

In this Beltaine-eve ceremony, held by moonlight, and enacted throughout the country, a solar-lunar true-to-all-life 'play' was re-enacted, in which parents and children mimed the sacred horse's part in their passage through (or over) the life-giving bonfire. Such rituals, undoubtedly prehistoric in origin, heralded the coming of the summer half as a supernatural event. From inside those horse shaped tunnels the stored materials, ready to make a prancing 'White Horse', may have been brought forth on Beltaine eve. A wooden frame, horse skull, and white sheet all quickly fitted around a dancer on the Chamber A threshold.[75]

With a single jump the flapping beast could then span the 7-feet-wide fire pit. After two more strides it could then leap over a nearby stone slab deeply engraved with concentric 'solar' circles, in a style introduced before 1000 B.C. This flagstone was found in 1929, lying on the courtyard's surface, scarcely 20 feet from the fire pit described above. The slab is decorated on both faces ('an extremely rare circumstance', in Macalister's assessment).[76]

The double display matches the sun's 'rise from the underworld', as did the fire, emerging from the adjacent pit. In addition, the first shaft of Beltaine sunlight, at 64 degrees from due north, passed over the causeway before striking the group of post-holes, arranged beside the Mare souterrain's womb. For centuries, these holes may well have housed a sequence of Maypoles, so providing the annual focus of sun blessed May-Day celebrations at ground level.

Though his Stallion souterrain is permanently fixed under Site 17, The Dagda is simultaneously ploughing his furrows between Ireland's north and south extremities, while also moving across the sky from dawn to dusk and back again.

When working to the northern end of this land-line, he could take shelter at Grian Ailech, Co. Donegal, a monument which he part owned.[77] It lies 170 kilometres from Uisnech. Despite the fact that to human eyes, the solar horse is never witnessed rising due north, he nevertheless shares the massive stone cashel of Grian Ailech with Grian, an equine sun goddess, who was believed to live within its thick stone walls. And there, at each day's end, The Dagda stabled himself with her.[78] Beyond this circular cashel, and on the same alignment from Uisneach, lies Malin Head, Ireland's storm-lashed most northern point.

Addressing the opposite extreme, and due south from Site 17, The Dagda's vapour trail passes directly over the Slieve Bloom mountains, incorporating Ard Eireann, a 526-metre-high peak, named after Eire, the island-wide deity, there seen in her maternal aspect.[79]

On the same alignment, and 112 kilometres from Uisneach, continuing his due south route, The Dagda can alight on Slievenaman, a 719-metre-high peak in Co. Tipperary.[80] There he encounters his underworld rival, King Midir of Bri Lieth, who had eloped to that summit from Tara with the 'Speedy Horse', Etain, the pair having flown there as swans![81]

So, thanks to overlapping myths, and the craft of metamorphosis, Ireland's Uisneach-centred topography incorporates natural features, monuments, stories and sky into a horse-powered living, repeatable whole, in time with the solar regime of night and day. And what could be more rational than that?

Festival of the Fires

In recent years, a Festival of the Fires has returned to Uisneach Hill at Beltaine. Attended by 4,000 people in 2011, this revival of the ancient oenach continues to herald the coming of the summer half of the year with music, displays of horsemanship, and parades of gigantic horse sculpts, around a vast central bonfire. In keeping with prehistoric tradition, this central blaze is answered by dozens of other fires, lit on mountain-tops, visible from Uisneach, in a nation-wide response.[82] On that night, the souterrain fire horses get plenty of exercise.

A Hindu Comparison

The Irish language belongs within the Indo-European family of tongues, and ideas about the horse are shared. In *The Bridaranyaka Upanishad* we read the following statement: 'The dawn is the head of the sacrificed horse. The sun is its eye; the wind is its breath. Universal fire is its mouth. The sky is its back; the atmosphere is its belly. The year is the horse.'[83]

At Site 17, the Maypoles were raised in Togherstown townland, which includes all the north end of Uisneach Hill; and since the name togher is probably derived from Irish *togha*, 'act of choosing, wish, desire', we might allow the Maypoles, combined with the land itself, to have the last words.[84]

LADY GODIVA OF COVENTRY

Reconnecting Humanity with Nature

Woven from several threads, the many aspects of Godiva's story make a happy mixture of incompatible elements. She avoids an either/or choice between past and present, while binding together 'fact' with 'fiction'. Consequently, she is never 'out of date'.

Her character is composed of Christian elements, themes drawn from European folklore, and traces of a Classical inheritance, including the Tyche, or city guardian female figure, here seen as old testament Eve, as a naked prototype, integrated into Coventry's local topography, and the celebration of seasonal renewal by a farming community in an agrarian hamlet, merging into a major industrial city. Godiva serves to create cohesion from diversity, while her current human representative is actively working to promote inter-racial and inter-faith harmony within the city.[1]

Named *Godgyfu,* meaning 'God's gift', the first Godiva was a woman born in the late 11th Century, but whose lands were not confiscated after 1066 and the Norman Conquest of Britain. Married to Leofric, Earl of Mercia, she was renowned for her beauty, generosity, and Christian piety. She founded and furnished

Coventry's Benedictine Abbey, in 1073, adorning it with gold and silver to an exceptional degree, according to medieval writers. She was particularly devoted to Mary, around the neck of whose effigy, she hung her necklace of precious stones, thereby identifying herself with the mother of Christ.[2]

To the all-but enslaved serfs of Coventry, a feudal village with 23 ploughs, Godiva became the living embodiment of a much needed god's mother, especially after Leofric imposed harsh taxes on them, including the *Heregeld* tax, to pay for the king's bodyguard.[3]

Lady Godiva had repeatedly, but without success, begged him to repeal that tax. In order to silence her, Leofric said that he would do so if she would ride through the streets of Coventry at noon, dressed only in her long hair.[4] To his amazement she accepted the challenge, and duly rode there naked, on horseback, as he had mischievously stipulated. Consequently the earl was obliged to remove the tax, and his wife was immortalised.[5]

Godiva died before A.D. 1100, but the first written account of her exploit did not appear until Roger of Wendover's *Flores Historiarum*[6] appeared two centuries later, and was confirmed by Matthew of Westminster's text of 1300,[7] in which he stated that she rode 'before thousands of spectators, *sans culottes*, clothed only in Chastity', and nakedness was a sin in the Middle Ages.

In folklore, similar tales are known from India to Norway[8] while Robert Graves regards the Godiva legend as a monkish adaptation of a pre-Christian ritual, involving the naked Goddess Nerthus' annual springtime rides across Germanic northern Europe, as described by Tacitus' 1st Century B.C. account.[9]

Underlying these was also the need to establish a better female-male balance in matters of law and social justice. Godiva came to stand for fairness, if not equity, in these areas; equally, she gave a focus to seasonal sympathetic magic, normal in farming communities. So in Coventry here was an annual procession to Cofa's Tree, a deep-rooted symbol of seasonal continuity, and a tree from which the name Coventry may have been derived.[10]

That tree also stood for the inescapable Life-Death, underground-surface double act. Wherever it stood within the village's 6,000 acres, Godiva's ride to its foliage helped to reconnect humanity to Nature, and thereby promote the possibility of finding Paradise on Earth. Cofa's Tree was presumably identified by the first Saxon settler, c. 500 A.D.[11] Today it rises nowhere yet everywhere in the tree-planted town, while lying horizontally among the war-demolished remains of her monastery, and rising again in the Cofa's Tree Paint Stripping Firm, operating at Unit 1, Spon End Industrial Estate, Coventry; a firm established 25 years ago, 'offering,' it says, 'friendly service'.[12]

An alternative origin for Coventry's name is provided by its early form Couaentree, which draws on the Anglo-Saxon word *Cune*, 'river confluence', since the Sherborne stream meets the Radford Brook immediately west of the city centre. The conjoined waterways are now culverted under Victorian Windsor Road. Together, in the 19th Century they were known as Arthur's river, and in the nearby Swans-Hill pool, it was said that King Arthur's treasures lay hidden.[13] When dredged and drained only mud was found.

Yet as a possible mistress of the waters Godiva's power may, with a long-shot, overlap with that of Coventina's, a Romano-British deity who presided over a richly endowed well and temple on Hadrian's Wall, Northumberland.[14]

Perhaps in Godiva, Tree of Life and Holy Well meet, to revive memories of the pagan Anglo-Saxon fertility goddess *Freyja*, widely known as The Lady.[15] Godiva's pagan aspect is also celebrated in the stained glass window of its Holy Trinity church. There a woman is depicted holding up a branch, as if promoting a spring festival.[16] Elsewhere in the city, several pre-Reformation panels illustrating 'The Labours of the Months' have survived.[17]

Yet following the Protestant Reformation, dramatic events, supported by music or visual imagery, were strongly discouraged, especially if they reflected delight in the Virgin Mary. But public commitment to Lady Godiva and her ride could not be entirely suppressed. Instead, the new churchmen decreed that *it should not be witnessed*.[18]

During 'The Ride', the population was told to 'remain indoors and with all their shutters closed', an absurd compromise that was tightly maintained for more than a century. However, by the late 17th Century, it was rumoured that Godiva's groom had bored holes in his shutters in ordered to watch his lady ride by, and that as a divine punishment for doing so, he was immediately blinded.

If this Peeping Tom incident introduced a salacious note into her story, akin to that of our sea-side 'saucy' postcards, Godiva's adult show nevertheless retained something of its full range of significance, even when, as in 1685, her part was plated by a boy.[19]

In the town's Godiva Day programme for 1809, outlined below, 'Peeping Tom' is represented by the early Tudor effigy of St George (England's patron saint). At other times, this 500-year-old wooden figure is displayed in a glass case, positioned to overlook Godiva's life-sized bronze equestrian statue, centrally sited in Broad Mead Square. Moreover, everyone is now invited to see what Tom saw, though in 1782, Godiva was required to wear close-fitting silk, flesh coloured clothes, and in 1811, her silk was replaced by a cotton garment. Today, no such compromise of true nudity is considered necessary.

Here is the order of the Grand Procession of the Show Fair for June 6th, 1809:

'The route will lead through: Hay Lane, Little Park Street, St John's Street, Much Park Street, where the fair will be proclaimed. Then on to Jordan-well, Gosford Street, Far Gosford Street, to High Street, where the Bablake Boys will sing. Then to Spon Street and West Orchard, where the Bablake Boys sing again; then Well Street, Bishop street, and Cross-cheaping, where the Bablake Boys sing yet again; then via High-street, returning through Hay-lane, to Trinity Church-yard.

> The order of the Procession
> Twelve Guards, two and two
> St George in armour
> Two Bugle Horns
> City Streamer
> Two city followers
> City Streamer
> Grand band of music belonging
> to the 14th Light Dragoons
> High Constable
> LADY GODIVA
> With city crier and beadle on each side

Mayor's Crier
City Baillifs
City Maces
Sword and Mace
Mayor's Followers
The Right Worshipful
The Mayor
Aldermen
Sherriff's Followers
Sherriff's common council
Chamberlains and Followers
Wardens and Followers
Grand Band of Music Belonging to the 1st Regiment
of the Warwickshire local Militia
Companies
Mercers, Streamer, Master and Followers
Drapers, Streamer, Master, and Followers
Clothiers, Streamer, Master and Followers,
Four Drums and Fife
Blacksmiths, Streamer, Master and Followers
Taylors, Streamer, Master and Followers
Cappers, Streamer, Master and Followers
Wool-combers, Streamer, Master and Followers
Butchers, Streamer, Master and Followers
Grand Band of Music
Shepherd and Shepherdesses
With a dog and lamb, etc.
Jason with golden fleece and drawn sword
Five wool sorters
Bishop Blaze
And Wool Combers
In their respective uniforms
Four fifes and Drums.

In this fashion, by accepting Godiva, the Civic Authority, the Church, and the military, along with the ranks of city tradesmen and rural shepherds, affirmed their interdependence and unity. But in 1845 mayor William Clerk suggested that the entire event 'that has for so long disgraced our city' be cancelled forever. His proposal was ignored.[20]

Godiva survived the carpet bombing of the Second World War, and is fully engrossed in the commercial life of modern Coventry. For example, The Coventry Building Society uses her image as their trade mark. She also appears by name in the city's Godiva Health, Godiva Carpets, Godiva Chiropodists, Godiva Dance Academy, Godiva Dental Lab, and Godiva Diamond Drilling. Her name also dignifies an estate agency, a fish shop, a glass supplier, a gun shop, a scaffold company, a school of motoring, a sign makers, a tailor, a wine merchant, an insurance company, a reptile and pet shop, a vet, a printer, and a dog-training school. Godiva, it seems, has become almost synonymous with Coventry. So, after the Coventry football team had been forced to migrate to Northampton for a year, after losing their home ground, on their return in 2014, they festooned Godiva's bronze statue with blue ribbon, their team's colour.

Both at home and abroad, Godiva is good for trade. Thus Draps Ltd, a luxury Belgian chocolate maker, launched their gold-wrapped Godiva brand, now on sale in 245 outlets across the USA. Meanwhile, since her discovery by E.L.G. Powell in 1982, she circles the world as Godiva no. 3018, a small asteroid in our planet's main asteroid belt.

Although Coventry is full of hard-headed entrepreneurs, the finer aspects of her enduring personality are not neglected. Since 1982, a Coventry woman, named Pru Porretta, has annually ridden the Godiva role, and has become the city's unofficial ambassador. On September 10th each year she organises a *Dame Goodyver's Day*, focusing on world peace and unity. Her Godiva Sisters include members of the city's now ethnically diverse community, representing different religious groups. In 2010, she was awarded an MBE, for her support of international charities. Porretta breathes the Godiva spirit, *and* she has long blonde hair.

Her Godiva predecessors include some London actresses and dancers, and a whore from Birmingham.

Yet the emergence of the feminine into public prominence continues

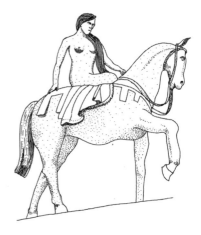

A bronze, full-sized statue of Lady Godiva on her horse, in Coventry.

to cause resentment in some quarters, and as recently as 1998 the city council again considered (but rejected) the idea of 'eliminating Godiva from the city's identity'. Fifty years earlier some councillors objected to her image being included in the town's central clock. Yet today, from that clock, she reappears, mounted on her horse, as every quarter hour strikes.

Moving with the times, in the 1911 procession which also marked the coronation of George V, she agreed to be accompanied by a *nun*, in order to quell anti-pagan protests. By contrast, her 1998 ride was linked to the centenary of the car's invention (Coventry has made a lot of cars), and in a 1994 version of her story, performed at the city's Belgrave theatre, she rode onto the stage on a motor bike, naked, naturally.

Because of her car industry links, combined with the town's central location within Britain, Godiva's horse has merged with a wheel. As a Godiva wheel she now features in university engineering faculties in Canada and the USA, where in both aspects she features in their annual festivities.

Godiva has also been adopted into the arts, having featured in Mascagni's opera *Isabeau*, and in two dozen pop songs. Six films have been based on her story, starting with the Vitagraph company production of 1911, followed by Gladys Jenning in the title role in 1928, and an unforgettable *Lady Godiva Rides Again* (1951), with Diana Dors in the saddle. Since then, three more Godiva films have appeared.

Coventry City's Herbert Art Gallery has a large collection of Godiva paintings, the oldest being a work of 1586, by the Flemish artist, Adam van Noort. Among poets, Ezra Pound and Sylvia Plath have both turned to Godiva for inspiration. One may truthfully say that she has infiltrated every branch of our culture.

In 1951, during the festival of Britain, Lady Godiva's procession was five miles long. It ended at the War Memorial Park[21], a reminder that the city has made great efforts to seek reconciliation with former enemies, in keeping with Godiva's maternal generosity of spirit, operating in effective deeds. She asks us to do more than merely peep.[22] Like a good steeple chaser, her horse leaps the boundaries of gender, prudery, class divisions and historical periods. From her central clock, she rides beyond time, and, with a world-wide reach, transforms Saxon *Godgyfu* into an active deity.

Lady Godiva's influence spread throughout the Midlands and beyond. In St Briavels, Gloucestershire, in 1820, the common land was saved for the people, when threatened by its planned enclosure, after Lady Hereford's naked ride around that village. The event suggests that a faint memory of the prehistoric Great Goddess might have underlain this incident and the entire Godiva persona.

Godiva frequently visited the small town of Southam, 13 miles south of Coventry.[23] There she led the annual Show Fair, during the 18th and 19th Centuries, centred around its holy well. Her companion was often a dark-skinned lady from Jamaica. A man wearing a horned bull's mask played a key part, as did a shepherd and shepherdess, with their lamb.[24]

And of course the well-known children's nursery rhyme, originating in Banbury, 30 miles from Coventry, urges us 'To ride a cock horse to Banbury Cross, To see a fine Lady upon a white horse, With rings on her fingers, and bells on her toes, She shall have music where ever she goes.'[25]

To this day, Banbury's version of Godiva is combined with the hobby-horse that features in British folklore. An annual parade and race of locally made hobby-horses is held in Banbury, overlooked by 'The fine lady' mounted on her bronze horse, sculpted by Denise Dutton and Artcycle Ltd in 2005. This confidently pagan work of art stands close to a 16-metre tall cross, erected in 1849 to replace the medieval Banbury Cross, torn down c. 1600. It celebrates the marriage of Albert to Victoria, with statues of her successors, Edward VII and George V, added later. These royal 'Peeping Toms' gaze across the road towards their mounted 'Lady Nature', as she rides on her endless journey.

VII.
Centres and Edges

SOME CENTRE POINTS

As Mircea Eliade's studies of mythic structures revealed, the cardinal directions, running north, south, east and west, correspond to the back, front, right hand and left hand of the human observer's body. The observer stands at the centre, the crossroads, where these parts intersect and combine, making a fifth location, namely 'Here'. Here too, the daily solar events of midnight, noon, sunrise and sunset are registered along the very same lines, thus linking the individual to a world-centred solar system, (regarded, prior to Copernicus, as the middle of the entire universe), and joining each individual to the starry cosmos beyond the Sun.

Moreover, the crossing-place of the cardinal directions becomes *the axis mundi*, or navel of the world, where a central tree joins the sky to the earthy underworld. Consequently, most versions of God-filled space tend to display the symmetries of the human worshipper's vertically standing frame.[1] A shared geometry was fundamental to worship, and so the cardinal directions were not regarded as 'abstract' terms. Rather, as recently as the 17th Century, it was customary in England for qualities to be linked to these directions. Thus, 'on a parish, hundred, and county level, The north stood for greatness, the east for health, the south for neatness, and the west for wealth', as William Lambarde reported in 1576, in his *Perambulation of Kent.*

The overarching experience of four cardinal points and a centre is found widely on every continent, and forms the basis of many cosmologies.[2] In ancient Ireland, the fifth province called *mide* (neck) was *here* to everyone in Ireland, and combined the distinct qualities of the four surrounding provinces. Furthermore, by means of an imagined umbilical string, Mide attached the entire country to the sustaining womb of the deities.[3]

Irish Mide is centred on the Stone of Divisions on Uisneach Hill, in Co. Westmeath, where a mighty five-metre-high boulder, (an Ice Age erratic), symbolically incorporates the four surrounding provinces into a single mass.

Likewise Pumlumon, the central mountain of Wales, is named after the Five Beacons around its summit, from which Wales is visible in its entirety. The beacons were Bronze Age artificial mounds, upon which bonfires were probably lit. On Pumlumon, a natural quartz effigy of the 'White Cow and her Calf' create an additional supernatural feature – the supreme bovine and her offspring stand available for all to see.[4]

In England the commonly used place name, Middleton, is often surrounded by one or more of the following village names: Norton, Sutton, Easton (or Aston), and Weston, as if to confirm the underlying fourfold north-south-east-west pattern, that legitimises the Middleton claim to centrality.[5] The centre is the point where irreconcilable opposites such as light and dark, or rising and setting, are found to be compatible.

Likewise, the relation of the individual to the tribe, and to the cosmos, is there confirmed. The centre is an incurable muddle of contradictions, which leads us towards an understanding of innate, existential, complexity.

Middleton also declares that the 'observer' is enmeshed in the 'result'. The scientific detachment urged on us (as obligatory) since the late 17th Century, is notably absent in Middleton; for there, by contrast, we apparently prefer to be incorporated into a world of attachments – a feminine outlook. The centre as *a noun* also involves the active process of *centring, and* in any language. For example, Welsh *canol,* 'centre', gives the verb *canolbwyntio,* 'to centre, to concentrate'.

In Britain there are 26 places named Middleton and 33 instances where the name Middleton precedes an extra term, such as 'On the Wold', or 'On the hill'. But how can this plethora of rival Middletons be justified? Are they as numerous as the number of founding inhabitants, each of whom is willing to state that he or she *is the* centre of a known world?

Some Middletons, such as Middlesbrough, North Yorkshire, have grown into a city. As for the range of outer settlements, reflecting the influence of a central Middleton, no complete set survives. Yet some 'Middles', such as Middle Littleton, Worcestershire, *are* flanked by a north and a south village, both sharing the 'Littleton' name. By contrast, in Hampshire Middle Wallop defines its

*An 18th Century view of the centre point of England
at High Cross, Leicestershire.*

satellite partners as Over and Nether, while Middle Winterslow in Salisbury, Wiltshire has the neighbouring settlements of East and West Winterslow in support.

Middle Aston, Oxfordshire, is echoed by a North Aston, and, to the south, by a heavenward, or noonday sun-searching, Steeple Aston; and another 'Steeple' village furnishes Middle Barton with a 'stone finger', rising skywards to its south.[6]

Yet no English 'Middle' has preserved a full set of cardinally directed neighbours. The ideal array has been smudged by time's carelessness. But the underlying 'four plus one pattern' continues to yield exchange and market day opportunities, and provides a framework for social gatherings, beyond the danger of inbreeding.

While every Middleton may be seen to flower, with extended petals, blooming across the world's surface, it is dependent on a stalk, rooted in the underworld, where some of the old gods live. In short, much centre-making comes from the original sacred dimension of human experience.

221

This is reflected in the division of Britain into parishes, formed more than 1,000 years ago. These were based on each Christian church's area of ministry, and the extent of the incumbent priest's responsibility, in connecting his flock to an underlying sacred reality.

In turn, the parish system often reflects the distribution of earlier pre-Christian sacred sites. Consequently, the pattern of pre-Christian worship was echoed in habitual village ceremonies, featuring pagan song and dance, which continued to be held in the churchyards. These ceremonies retained something of their identity as pre-Christian sanctuaries, additionally enhanced by the veneration often accorded to an old churchyard yew tree, which predates the church. Given this double load, the knolls on which so many parish churches are sited, suggest that the Earth itself is rising to meet the need.

In Wales, at least 380 place names begin with *Llan*, 'church', firmly defining all secular structures as derivatives of sacred reality. While in England more than 300 settlements are still named after saints. Among these, saints are Bride, Ann, Helen, Margaret and Mary, who all feature repeatedly. The influence of a holy individual from the past is invoked in order to insure the centre's living worth, by nominally including one of the first and best of the believers among the flock of later devotees.

The process of centring also involves time; the accumulation of a densely layered time, made possible by means of habitually repeated custom. In E.P. Thompson's words:

> 'Custom was ambience. It served to structure lived environment of unwritten beliefs, and sociological norms. It was the foundation of a world informed by oral tradition and expressed by repeated exercise,'[7] (such as annually carrying Swallowhead spring water to the top of Silbury Hill). 'Custom pervaded the routines of livelihood with the ambience of tradition. Practices were allowed because they had been anciently used; rituals were enacted because precedent dictated. So the understanding of the past informed and infused the working of the present'.

Equipped with, and centred upon, such a multi-layered

understanding of time, people could prophesy into the future with some confidence. 'The natural world was full of prognostics and prophetical charms, which women seemed to have specialised in perpetuating down the generations'[8], partly by means of their fireside story-telling on winter evenings.

So, each Middleton defied the dictates of a strictly linear chronology, and around every central hearth, Middleton enjoyed an all-time abundance.

THE MIDDLETON BREAD MAN, OR CHRIST NOT SEEN DEAD IN A SUPERMARKET.

As an artist employed by the Rochdale local authority in 1989, I was asked to make a sculpture, to be sited in a small rubble-strewn plot, near the town centre of Middleton, Lancashire. Two of the plot's sides were defined by the high, windowless walls of a multi-story car park, and the equally windowless side of a supermarket. Both these structures were built in a 1960s modernist idiom. The hypotenuse was formed by a main road; along its pavement customers pushed their loaded trollies from supermarket to car park; thereby completing a Food Triangle. So food became the obvious subject for the sculpture, which I was instructed to place against the car park's brick wall. Further signs of food blew across the site – a plethora of chocolate bar wrappers and discarded crisp packets, which sometimes danced together in a spiralling eddy, over the bare earth floor, while a cluster of ripe elderberries trembled on a solitary self-planted shrub.

The elder, neither bush nor tree, is *Sambucus Nigra*, the embodiment of the Elder spirit, or elder mother, in British Folk belief. Judas is said to have hung himself on it. In some accounts so did Christ. The plant's white blossom, black fruit, and stinking boughs, brittle as dry bones, help to sustain a powerful mixture of associations.

Hawthorn and bramble were also beginning to colonise the ground, attracting insects, preyed on by spiders, in turn picked at by birds; which were at times watched with appetite by a stray ginger cat. Together they made a work of non-human food art – an all-devouring round dance, in which participation was compulsory and included the nearby human animals, who naturally sang to the same tune, with the same unending 'I eat you; you eat me' lyric.

What was already happening in this triangular wilderness at the centre of Middleton could therefore be described as 'The Heart of Art', the background rhythm around which the achievements of world culture arose and arise; a pulse, audible when Middleton, as an Anglian village, was established, 1,500 years previously.

The artwork for the site could take the form of a passive acceptance of this situation, leaving it to chance whether or not a pedestrian might one day gather and eat the elderberries, snare the birds, or try to improve by selection the wild grass heads, planting them in rows for greater convenience, to repeat what happened here in the early Neolithic, c. 5000 B.C.

But then the Parks Department spread topsoil across the site, demonstrating that absolute wilderness is hard to find. Turning my attention to the multi-story car park wall, its unpromising exterior was enlivened by Le Corbusier's remark of 1955: 'One day, passing at the foot of that wall, behind which the gods are at play, I stopped to listen.' To his insight, I could add that of his supporter, Geoffrey Scott: 'The naïve, the anthropomorphic way which *humanises* the world, and interprets it by analogy with our own wills, is still the aesthetic way; it is the basis of poetry, and the foundation of architecture.'

So perhaps the Middleton wall that I faced was the very one, the first to be measured by the six-feet-tall fictional English hero, Bulldog Drummond, whose 72-inch height was to provide the basis of the entire Corbusian modular system. And, 'yes', this wall *was* precisely four Bulldog Drummonds high: 72 inches times $4 = 24$ feet.

Moreover, between the concrete of the post and beam construction, the brick infill panels presented an array of baked clay objects – bricks, each 9 inches (22.8 cm) long, resembling both the racks of cooked loaves in the food store, and less obviously, the parked cars arranged in tiers, behind the wall, each awaiting, and soon to be loaded with one or more of the shop's loaves. This coincidence brought *the staple food-*

item of our diet into focus. Underneath the surface litter of crisp packets and discarded Mars Bar wrappers, lay bread, essentially the same loaf that was grown by the first Middleton farmers, and eaten at Christ's Last Supper.

The food theme was also steered further towards bread by the walls, four load-bearing concrete uprights, painted scarlet by the artist at an early stage, in deference to the theory of 'functionalism' that they supported. So painted, the pillars accidentally defined the three panelled shape of a Christian triptych, and thus brought with it the prospect of a holy bread delivery.

The Proposed Loaves
I decided to make a set of double life-sized fibreglass loaves to fix onto the car-park wall, to be appreciated for their differences of shape, and for their ability to come together in new/old patterns of meaning. (The types of loaf involved here were also common in the south Midlands, where the artist had done a Saturday Bread Round in the early 1950s.) Here they would be arranged to make the image of one Giant Bread Man. The 'Bread-man's' body would be composed of the following parts:

The Feet: French loaves, baked a deep orange-brown. Shaped like 15th Century foot armour, but twice as hard, and terminating in mean snouts, that can explode when sawn. A major cause of bleeding gums in Buckinghamshire.

The Shins: Split tins.

The Thighs: Two Bloomers; the bloomer is a complex loaf. Its long, baggy shape is deceptively casual, and wounded by parallel diagonal cuts. Its base is hard, flat and floury, yet a line of blond tenderness often runs between base and upper arch.

The Loins: For decency's sake, the loins to be girded in an enlarged version of a 'Sun Blest' bread wrapper, also made of fibreglass, and involving a pun on the Jehovah-Christ connection.

The Stomach: a Coberg loaf, that has a cross cut at the centre, suggesting the four quarters of the ideal town plan, (just as the name Middleton implies the surrounding villages of Norton, Sutton, Easton and Weston); and also the umbilicus, or navel of the world.

The Bread Man, a sculpture by Michael Dames, at Middleton, Lancashire, 1980.

The Chest: Tins, the standard 'thunder cloud' shape, swelling above the tin's rim, to a dark brown (often black) convexity, sometimes clinging stickily to the teeth, when eaten.

The Upper Arms: Split tins.

The Lower Arms; Baguettes.

The Hands: Baps.

The Head: a Cottage; a double decker, two tier loaf, with the smaller of two rounded forms perched on top.

When all securely bolted into the wall, the loaves might collectively display a sacrificial Christ crucified effigy, coupled with a Last Supper, bread-only event. As for the overlapping references, there is no need to invent multiple perspectives, for they already exist in profusion. They constitute normality. Therefore this was a truly *realistic* artwork, and when simply delivered, it would the better reveal the varied and compound origins of its simple form (just as a broadly polished rock face will show both the varied origins of its components and an overall unity).

What H.G. Gadamer termed *Verstehen*, 'a mingling of horizons', often occurs in art-making, revealing the compound nature of most present-day surfaces. For as Gadamer claimed: 'There is no horizon of the present for itself, nor any isolated historical layers, but always a merging of these horizons'; even where, as in Middleton, it transpired, most people prefer clearly demarked, single purpose signs.

I decided to paint the flanking brick panels in the gold and white undulations of windblown ripe corn fields, to allow the loaves a glimpse of their pre-baking earth-rooted history, and to suggest a possible country outing for the town-locked parked cars.

In deference to Middleton's original pre-Christian inheritance, I decided to flank the Christ-Commerce central figure with two 12 feet (3.7 metre) tall 'corn dolly' effigies. These were made of aluminium scaffold tubes, and modelled on the traditional shapes of (1) a Mother Earth and (2) a Hag or Maiden corn dolly type.

Such Goddess deities were annually plaited from the last straws of one harvest, and ploughed back into the land at the start of the next ploughing season. So in many examples, such as the Herefordshire fan dolly, the dolly is known as both 'hag' *and* 'maiden'. Here in giant metal form, they would combine rural and urban references: while as pagan 'idols', or 'dollies', they merge with a 'Christian' triptych and secular shopping experience, in which food may nevertheless still be involved with a search for the meanings of life and death.

Given these corn dolly companions, Christ can also register as John Barleycorn, perpetually rising from death in the mill; here represented by the pit, excavated below him, where a millwheel is set to turn perpetually.

Given the strength of the Protestant Calvinistic tradition in the area, as a matter of courtesy I visited all the clerics in the district, before beginning the work. None raised any objection, but after it was completed, a fundamentalist group, based in Manchester, raised a petition to have the installation removed. Some people objected to the smiling Sun-Blest logo on the loincloth, and although the manager of that bakery wrote to the *Middleton Guardian* disclaiming any connection with the effigy, the completely untrue rumour persisted that I had benefited financially from the connection. What I was doing was to raise questions about the interplay between secular and sacred aspects of reality, since Advertising Man and God as Man, are bound to share the planet, while the sun – even in rain-soaked Manchester – shines on all, gathering the differences into a 'being-at-one'. To be human may involve accepting the whole cultural deposit, taking pleasure in its variety and richness.

Yet among the 1,200 petitioners demanding the removal of the work, many found it 'trivial or profane, an insulting monstrosity… an abomination… sordid, horrible, repellent and disgusting'; while letters in support found it 'original, thought-provoking… a cause for joy; and Jesus would be glad of the publicity'. In December 1980, Councillor Hunt, in rejecting the petition, said: 'It is here to stay, and most people don't take the slightest notice of it.'

Curiously, the only part of the ensemble that was not publically vilified was the altar to the Mother Goddess, located beneath her metal skirt. This could be seen as a special kind of loaf, an up-rearing of Earth, the definitive bun in the oven, baked by time, offering the annual first fruit, and a measure of all subsequent loaves. It was, 'A token of the Blessed things that have taken on what is cursed, and the profane things that are somehow redeemed,' in David Jones' words. By 1991, the entire ensemble had been removed and the walls repainted a dark red, the colour of dry blood.

In retrospect, I see the departed Bread Man sculptures as an adventure into the normal condition of the post-modern world, where apparently irreconcilable outlooks are asked to contribute, as in a collage, to a unified whole, with their horizons intermingled to make a new reality from old scraps of material, just as a variety of wild flowers can help to realize the overall beauty of a field.

Whereas 'History' improbably requires us to believe that we live on one page at a time, with the old outlook being totally replaced by the next sheet, *in fact* different things and different outlooks accumulate, and interpenetrate. So in this work, I brought together the Christian religion and today's shopping (surely if Christ is alive today, he might be seen in a supermarket), and town and country (corn dollies and modern architecture; Middleton began life as a rural, corn growing community).

Hence straw turns to aluminium, in front of a painted cornfield, with mother, maiden and hag as different faces of an annual cycle. I also included the underworld, the land surface and a smiling sky as a possible heaven, with this trio implying a divine death and resurrection, hinting at an endlessly repeatable Last Supper – offering you your next sandwich and a glimpse of Eternity.

MYTHS OF THE LOWER SEVERN

Britain is an untidy island. Its surface carries detritus, accumulated from beliefs long since abandoned. These obsolete fragments match neither Christianity's disdain for the world, nor the modern landscape of de-sanctified materialism. Discarded myths can still offer us peep holes into former visions of unity. Such glimpses remain valuable today, for they can extend our narrow culture, and provide cyclical alternatives to its present unstable linear trajectory.

The Severn, like many other British rivers, timelessly reunites such cultural fragments as it runs endlessly from source to sea.

Naming the Severn

The Severn river's history is inter-woven with an abundance of myths. From its source at 610 metres (2,001 ft) above sea level, on Pumlumon, the highest mountain in mid Wales, the river eventually flows into the Bristol Channel, after a journey of 220 miles, making it Britain's longest waterway. But that says nothing about its nature. So what is the Severn?

Is it an inverted water tree, fed by, rather than giving forth, many branches, as tributaries? Or a lifeline, offering a home to innumerable competing species, including humanity? Is it a pathway for migrating birds? A major artery-cum-vein, to an island, regarded as a living entity, that feeds the main stream with innumerable capillaries? Is it also a story line, to which every generation has contributed its own twists and turns, while sometimes rewriting the entire plot? It is all of these things.

As a liquid 'motor way', it can be used as a badly designed canal, and as an inefficient drain. It also can serve as a mirror held up to sun, moon, stars and clouds, while simultaneously acting as a knife that, despite its bluntness, excavates its own valley. In doing so, it provides the underlying pulse for the river basin's range of poetry.

The Severn can also be viewed by merchants and explorers as a route to the world ocean. For those who wish to see her presently idiosyncratic route, determined by northern ice sheet blockage, she is a misdirected Goddess, who should be flowing into the Dee (Latin *Dea*, 'goddess') at Chester.[1] Or is she the princess Sabrina of 12th Century imagining?

Today the Severn serves as a convenient industrial sewer, doubling as a farmer's irrigation opportunity during periods of drought, while also offering a source of barrage-produced electrical power, in the future.

The Severn's original divine name, Sabrina, has been known since prehistoric times, when the female supernatural virtue of all Britain's rivers was recognised. Sabrina is synonymous with Britain's longest stream. She was acknowledged by the Classical writers Tacitus and Ptolemy.[2]

Her name finds an echo in the Irish River Sabrann of Co. Cork.[3] Both may contain meaning derived from Sanskrit *sabala*, 'the spotted cow of plenty, identified with the powers of Nature'.[4] (Morris-Jones prefers to invoke the Sanskrit root *sam*, 'happiness, welfare, and prosperity').[5] Most scholars agree that Sabrina comes from the Indo-European language family that has meandered like an ocean current around half the world.

Since the 6th Century A.D. Sabrina has shared her waters between England and Wales. In the latter country she is named Hafren. She started out as Welsh *Sabre;* this changed to Habren by A.D. 800 and to Hafren by 1150.[6] As a Welsh word, *hafren* can now mean an 'untidy, slatternly woman' – a description that fits Severn's tendency to burst her banks. In doing so she reoccupies the river's broad valley, (itself created by her own erosive power), to annually smear it with a fresh load of her deposited mud. In addition W. *hafren* can also signify a 'prostitute'[7]; perhaps because for centuries after A.D. 577 the river formed the shared frontier between enemies, when the Saxon Hwicce tribe defeated the British and settled on Sabrina's eastern bank, including around the ruins of Roman Glevum (modern Gloucester).[8] To the Welsh, it seemed that their water goddess would now go with anyone!

Yet in this respect Anglo-Welsh history was re-emphasising an underlying mythic pattern, in which a territorial 'Lady' (such as the Irish goddess Medb or Maeve) typically gave herself to every king of her choice, one after another.[9] Likewise Hafren's life-giving flow undermines any notion of individual permanence. She is faithful only to change.

From source to sea, (*Mor Hafren*, as the Welsh term the Bristol Channel),[10] Sabrina was seen as endlessly re-enacting a cycle of birth, youth, maturity, old age and death – terms now employed by modern geomorphologists in their studies of rivers. Yet within that remorseless process, shared by all living things, the Severn seems to emphasise a positive climax, since *hafren* contains Welsh *haf,* 'summer', and therefore the prospect of abundance. Likewise the name Sabrina (from which English 'Severn' is derived, via 11th Century 'Saberna') comes, according to Ekwall, from Sanskrit *sabar-dhuk*, 'milk'. If so, Hafren's harvest and Severn's 'milk' feed us across the language divide.

In this cycle, the East-West axis, defined by the solar equinoxes in March and September, when night and day are of equal length, was emphasised in Welsh Severn-side myths, by the story of Bendigeidfran, son of Llyr. He was imprisoned on Gwales, (modern Grassholm), an island lying off a Western tip of Pembrokeshire. After breaking a taboo forbidding him from looking south across the Severn towards Cornwall, his decapitated head was carried east and buried under the Tower of London,[11] thus establishing an East-West axis more than 150 miles long and spanning almost the entire island, which links solar and kingly energies, and hints at Welsh recovery of an island-wide control. Coupled with the eastern burial of Bendigeidfran's head, buried underground in London, the ancient Welsh Triad, no. 52, lists 'Three Exalted Prisoners of the Isle of Britain'[12] who are incarcerated at different places along an 118-mile stretch of the lower Severn and its estuary, between Gloucester and Lundy Island.

Gwair, Son of Geirioedd

Lundy, a Viking name meaning 'Puffin Island', was (and still is) termed Ynys Wair in Welsh, after a youth, Gwair or Gweir, believed shut therein.[13] The isle is shaped like a sea serpent, with an open mouth facing south. It is a flat-topped granite rock, three miles long, extruded from the Earth's molten core, and walled by sheer cliffs up to 400 feet (122 metres) high, making it a natural 'fort' (Welsh *caer*).

Situated 31 miles from South Wales, and only twelve from Hartland Point, North Devon, Ynys Weir is thrashed by Atlantic gales that rip off herbage and soil, down to the bare rock in places. At other times the isle vanishes under fog that can linger for weeks. Yet in Celtic belief such an inhospitable western isle was always likely to be regarded as an Other World siddi, (Irish *sidh* – or *sid*), home of the immortal gods. Many experts regard Gwair as a version of Pwyll, who had stayed for a year in Annwn, beneath Glyn Cuch.[14]

From a mundane perspective Lundy plays many disturbing tricks, as J.R. Chanter reported in 1871.[15] 'Often a light vaporous mist magnifies objects a few yards distant to three times their actual size. A phenomenon akin to mirage also occurs whereby the outline of the island [seen from the mainland]… appears lifted and distorted… with ships sailing under it.'

With Virgin Spring named at the north end, and Mermaid's Hole at the south, Lundy's uncanny reputation was further enhanced by the Jacobean poet Michael Drayton, who described rival choirs of English and Welsh river nymphs, singing for possession of the isle. This idea drew on the folk belief that pipers and harpists played in the Other World, so sustaining the entire cosmos as a musical event, to be augmented by human song and dance. Drayton went on to personify Lundy as one of 'Soveraigne Severne's' estuarine goddesses 'that bathe their amorous breasts within her secret Deepes'.[16]

Even John Stowe, a sober Elizabethan historian, credited the isle with improbably sumptuous qualities, garnered from its supernatural Celtic endowment. He wrote: 'It aboundeth altogether with victualles, and is verie full of wines, oile, hony, corne, bragget, salt fish, flesh, and sea or earth coales'.[17]

Trapped inside Ynys Wair's *sidh* or *siddi*, the divine Gwair found himself surrounded by the riches of Annwn, the Welsh Under World, as the medieval poem *Preiddeu Annwn* states: 'Complete was the prison of Gwair in Caer Siddi... No one before him went into it. The heavy blue chain held the faithful youth, And before the spoils of Annwn dolefully he chanted.'[18]

As a god, Gwair is necessarily the first to be tethered within Lundy, pre-dating even the Mesolithic visitors who left their flints on the island c. 6000 B.C. He was obliged to inhabit and energise this supernatural nucleus from the outset because it served as a prototype from which the world, as known to humanity, is a secondary derivative. Therefore he endured Ynys Wair's confinement amongst the delicious foods and treasures of Annwn's primal store. Perhaps he was doleful because he found the 'spoils of Annwn' monotonous in their perfection. They epitomised the divine quintessence of all (and only) the good things that might eventually arrive, if in a diminished form, on Earth.

But his human aspect longed for the changes and hazards that real life brings. The 'heavy blue chain' that held him may refer (as in many other creation myths) to the encircling firmament, embracing the abyss. On a local scale the same chain re-appears as the ring of waves, pounding against the Lundy cliffs, and is also wrapped around every mammal, confined in prenatal 'submarine' protection, within the salty waters of the womb. Further, Ynys Weir mirrors Britannia's state, since she, too, is only a small patch of land, anchored on the planet's mainly liquid surface, into which Sabrina-Hafren pours fresh torrents, garnered from evaporated seawater falling as rain.

So, among other things, Gwair's immortal chain predicts Sabrina's role in the cycle of life, operating on both vertical and horizontal planes, while also joining together supernatural and secular accounts of reality. Involved as he is in such a demanding task, should we be surprised that Gwair 'dolefully chanted', or that the Welsh noun *gwair* can also mean 'bend, circle, loop', and 'collar'? He collars everyone, including Ynys Wair's Bronze Age inhabitants, who lived in round huts (up to 9 metres in diameter) – some grouped within a rough stone pound at the island's north end,[19] and the 20,000 day visitors, arriving each summer on the Ilfracombe ferry.

As a mythic character Gwair demonstrates that we are all caught up in a circulation of matter that over-rides artificial boundaries. He teaches this by example. Accordingly his very name cancels the distinction between sacred food and mere vegetation, for Gwair is also Welsh *gwair*, 'hay' – ordinary grass, cut, raked, dried, and stacked in midsummer. The swish of a haymaker's scythe finally releases him from rooted imprisonment in the ground. The largely pastoral economy of Wales depended (and still depends) on this, the hay god's yearly self-sacrifice. He rises, stands and falls to advertise the extraordinary origin and abiding worth of the commonplace. His presence guarantees ultimate quality within gross quantity. Triad 52 describes this same Gwair as 'son of Geirioedd', a name based on W. *geirio,* 'to word, to phrase',[20] indicating a talkative disposition; (Gods are often named according to what they can do). But if Gwair expected to inherit his father's power of speech, enabling him to utter and transmit his father's myth, he had to be patient. Eloquence would not sprout from Geirioedd's dry seeds until they had been given time to germinate. Therefore Gwair was imprisoned in silence for many months, within Lundy's thin, acid soil. Throughout the winter, he tried in vain to pierce the dark earth with sharp green blades that he could, at that stage, only imagine. Yet as Rachel Bromwich believes, Gwair's name stems from a root *veg*, 'awake, lively, stirring'.[21] Accordingly in time, the tips of his spears do cut through the gloom, into daylight. He then acquires and lives up to yet further meanings of the Welsh word *gwair*: 'fresh, sprouting, wanton, ardent'.[22] Therefore by May's Beltaine festival, when the Celtic summer half of the year begins, Gwair is once more standing at least a foot high, spraying pollen, while gesticulating in the wind, an all but silent orator, miming his father Geirioedd's vocal example. The yearning to find the origins of speech within Nature is surely reasonable, for where else can speech have come from?

Munched noisily in the horses' mouth, and then passed into its stomach and bloodstream, the hay god's 'spirit' changes from vegetable to animal form; digested, not lost, prior to its delivery as a fertilising excretion back into the field. Here physiological fact and poetic metaphor combine in pagan (rural) experience akin to the 'ground-of-being', where mind, heart and gut can together recognise the entire cosmic metabolism in a stem of grass that pipes 'antic hay' across the universe.

In 1798, a few miles east along the English coast from Lundy, the poet Coleridge tried to re-enter Gwair's reality, searching for some 'co-naturalness' between the individual mind or soul and Nature. He later wrote: 'A Poet's Heart and Intellect should be …intimately combined and unified with the great appearances in Nature'.[23] For many readers, Coleridge's *Ancient Mariner* poem met that need. Meanwhile on the banks of the Wye his friend Wordsworth felt, 'A motion and a spirit, that impels All thinking things, all objects of all thought, And rolls through all things'.[24] He, too, was riding Gwair's 'circles and loops', both great and small.

Gull, Gannet, Hen and Chickens, Seals, Dead Cow, Goat, Rat and Mouse now feature in the place-names of Lundy, along with Jenny, Ladies, Widow, Virgin and Mermaid; but Gwair is absent. He has been replaced by: Frenchman, Devil's Slide, Devil's Chimney, St John's Well, St John's Stone, and other Stones attributed to St Peter, St James and St Mark. Except for a few small hay fields at the southeast corner of the island, most of the pasture is now moor land or 'rough grazing'. The real Lundy falls well short of any pre-Christian Paradise, yet continues to point its granite rocks towards a notional ideal.

A Threshold Myth of Llyr Half-Speech

In her myths, as in topographical fact, river-sea relations are central to Hafren-Sabrina's tale. The moon-stoked 41-feet tidal range displayed around Chepstow, where the river Wye joins the Severn, is the second highest in the world.[25] There seawater surges up the rapidly narrowing funnel between South Wales and the Devon-Somerset shore, and piles up between the facing headlands of Beachley and Aust, set only a mile apart, on the Welsh and English sides respectively. Between them, in mid-channel, exposed to view at low tide, lies Leary Rock, the possible 'gaol' of another of the three prisoners listed in Triad 52: 'Llyr Half-Speech, who was imprisoned by Euroswydd'.[26]

Leary Rock is a mid-channel, entirely natural limestone reef. Lying in the Severn about 600 yards (549 metres) upstream from the former Aust–to–Beachley ferry crossing, Leary is now overlooked by the suspension bridge built in the 1960s. This flat slab of limestone measures over 300 yards (274 metres) from west to east, by 60 yards (55 metres) wide. The obsolete English word 'leary' meant 'pitted, as in a sea-cave,'[27] and matches the crevices in Leary Rock, into which tides swill. Equally, the name could be an Anglicised version of Welsh *llyr*, a common noun denoting 'the sea'.[28] If so, in Welsh eyes Leary Rock might appear as a gigantic effigy of their sea god Llyr (and also of his Irish equivalent Mannanan mac Lir).[29] This solidified 'Llyr' image lies midstream, like a basking Neptune, sculpted at the edges by Hafren's currents. His 'head' faces west, and his two shanks extend towards the English side. Imprisoned in (and as) the Rock, he lies close to where the riverbanks start to diverge into their widespread estuarine shores. Consequently, at Leary Rock Hafren-Sabrina receives the sea god's funnelled saline fluid in full measure and at speeds faster than any other British tide. The intertwined turbulence of sea and river is very intense, with the Rock positioned as a hazard to shipping and as a convenient marriage bed for the union of salt and fresh waters. Across its surface, millions of elvers (infant eels) wriggle upstream like spermatozoa; over a superman-shaped doorstep, washed by every incoming tide, that crosses the river's threshold.

In mythic affairs, place-names, repeatable events, divine names, words, living creatures and mineral 'things' are prone to coalesce, and here people may have heard Llyr's 'half-speech' in the hiss and slurp of in-rushing tides, though *what* he was saying could not be understood.

Indeed, when, at high tide, the marine god arrived to cover over Leary, his gigantic limestone self-portrait was then least visible, being wholly submerged by his inrushing waters. It was Euroswydd, his Welsh *euroswydd,* or 'mighty enemy', who condemned him to this *submarine* self-effacement, from where his foaming cries made an enraged 'half-speech'.

As to the identity of his Mighty Enemy, perhaps derived from W. *eur,* 'gold' plus *oswydd* 'enemy'; was he disparaging the entire solar system, including the sunlight that evaporated water from the ocean, which returned to earth as rain? Failing to see this connection, Llyr identified his immediate gaoler as the Queen of Tides, alias Roman Luna, Welsh Leuadd, and the English Moon. Unaware that without the ocean, (his inrushing tormentor), there would be no life on Earth; no rain to feed into rivers and fields. Instead, he would preside, dry as a bone, over nothing but rock. Instead, the unfortunate Llyr served as a male midwife to the entire Severn valley life-event.

Aust's Old Passage

For many centuries, the main route from England to South Wales involved a ferry boat journey the length of which was variously estimated at between one and three miles,[30] across the Severn estuary between Aust Cliff and Beachley in the Forest of Dean, a route known as The Old Passage. People shared the craft with horses, cattle, and other livestock. After considering 'the formidable fame of the tides and the Bore, and the winds that make the water so rough, and the boats so mean, none of us cared to venture,' remarked the intrepid 18th Century traveller, Daniel Defoe.[31] He and his party preferred to retreat to Gloucester, where they crossed the river by bridge.

The Old Passage threshold into Western Britain had long been revered. The prehistoric bronze image of a moon goddess, found at the base of Aust cliff in 1903, had been made at Despeñaperros in southern Spain,[32] and donated by mariners grateful to have made landfall here. It was one of a number of bronze images found at Aust,[33] but now lost.

In addition, a 1st Century Roman dupondius coin of the emperor Claudius also fell onto the foreshore at the foot of the cliff.[34] On its reverse, there was an enthroned image of the goddess Ceres. She was depicted carrying corn stalks in one hand, and a long torch in the other. With this she could search the underworld for her daughter Proserpine, imprisoned there by Hades.[35] The name Aust probably derives from Claudius' reverence for Augusta, his grandmother, whom he raised to the status of a divinity.[36] Aust Cliff, although a mere 134 feet (41 metres) high, is a divine eminence. (Christian writers have attempted to link its name to an oak, beneath which St Augustine conferred with the Welsh bishops).[37] So another layer of sanctity accumulated there, during the struggle to establish a fixed date for Easter, which has yet to be achieved.

In geological terms, the hill is a rich layer cake, reaching from a 240 million year old base, through the Triassic desert period, including its Rhaetic layers. These were named after the Greek giant Rhae, and the blood spilling from his severed testicles, and then what geologists term the tea green marl, overlain by a 140-million-year-old layer of Lias and the Cotham marble deposit, with its black, tree-like markings, and the famous 'bone bed', containing teeth of marine dinosaurs, one of which adapted its four flippers into legs, on which it made the transition from marine to land dweller.[38] The entire ensemble is now capped by modern soil, into which the motorway service station has been grounded since the 1960s.

Other species found there include one that resembles the lung fish, which now lives only in the antipodes.[39] Aust presents a museum of lost lives and faded beliefs, which can be revived within ancient texts, like Ovid's *Metamorphoses,* in which we are reminded that 'Ceres was the first to turn the glebe with the hooked plough share, and first gave corn, a kindly substance, to the world. She must be the substance of my song.'[40] It was partly their need for British corn that prompted the Roman invasion of Britain.

Aust is a hill-full of life, turned to artefact, and fossil-rich sediment, all advertised as holy by its place-name, implying a safe place for a river crossing, in any era.

Yet the Aust trip was undoubtedly hazardous. A boat was lost in 1839, with all passengers drowned, and another upturned in 1855, with few survivors.[41] From one sunk craft, seven passengers saved themselves by clinging to the tails of cattle, which had shared the journey with them.[42] In 1777 a man capsized a ferry when struggling to recover his hat, which had blown off in a gust of wind. All but one man drowned. His hat later washed ashore, its brim stuffed with bank notes.[43] The ferry boats were flat bottomed, and sometimes took up to 20 attempts to come alongside, given the strength of the tides. A smoke signal of flaming straw was lit at Aust, to call a ferry over from Beachley.

By 1830, six stage coaches a day arrived at Beachley, with the passengers crowding onto a wooden paddle steamer, named Worcester, built in 1827.[44] (It, in turn, was replaced by a 60-ton, iron vessel, with a 3-foot (0.9-metre) draught, in 1838.) From the 1920s, ships, such as the Severn Queen and the Severn King, operated as car ferries, until the nearby suspension bridge opened in 1966.[45] In 1752, the ferry charged 2/6 (30 pence) a score for hogs, and 4 pence per man, woman and child. All manner of living things shared these flat bottomed craft.[46]

At Beachley, all ships going upstream were obliged to put in, to be searched by customs men. At Beachley Point, on a rock only accessible at low tide, there was a navigation light and the ruins of a chapel, once inhabited by the Welsh St Twrog, (English St Treacle).[47]

Tirthas as Sacred Crossings
To understand a river's ups and downs, it is also necessary to travel across its line of flow, for a river would be nothing without its banks. As for this particular waterway, it had been a highway since early Neolithic times, and people have left their footprints along its muddy margins since the Mesolithic era, 8,000 years ago.[48] At Old Crossing Sabrina makes her transition from river, to estuary and then to the world's ocean, on which sailed Sri Chimnoy in 1991. He was beginning his world harmony run through 40 countries, as part of his world-wide dream of peace, during which journey on November 15th, he renamed the Severn-Wye bridge as the Sri Chimnoy Peace Bridge.[49]

In doing so, he evoked the tradition of *Tirthas,* or Sacred Crossings, that knit together India's Sacred Geography, in which rituals and reverence are linked to hilltops and rivers. The place itself is the primary focus of devotion, and can be linked to places of pilgrimage touching the shore of the Other World from the depths of the traveller's heart.[50]

Within a numinous landscape, places of pilgrimage are charged with value and significance, crossing linguistic, sectarian and regional boundaries.

Similarly around Old Crossing, Welsh creation myths speak of Henwen, a white sow from the Other World.[51] She emerged from a spectacular, vaulted round barrow, sited at Ballowal, close to Cape Cornwall, (which was, until recently regarded as that county's outermost tip). From there she journeyed up Mor Hafren, the Bristol Channel, and landed at Aust Cliff, then regarded as greater Cornwall's easterly tip.

And, as the medieval welsh *Triad number 26*, declares, 'from Penrhyn Austyn' (Aust Cliff), she 'went into the sea', with Col, son of Collfrewy, clinging to her bristles wherever she went.[52] Since his name in Welsh is 'beard of corn, husk and sharp like a sting', he serves as the protective sheaf of the corn seed that she carries, and also arms the first bee against any attackers. As the beard, on the soon to be born wheat, Col is accepted if not welcomed by Henwen as a necessary passenger.

After their trans-river dip, Henwen came to land on the shore of Wales at Aber Tarogi, in Gwent.[52] But she was no ordinary hog, but a supernatural sow, white, with red ears. She had emerged from blind Dallwyn's Other World domain.[53] The Tarogi brook, named from Welsh *tarrawg*, pregnant, joined the Severn at Southbrook Iron Age camp.[54] From there Henwen swam six miles upstream to give birth to the world's first grain of wheat and its first honey bee, at Maes Gwenith, Gwent – the same location where she had been born, according to Welsh belief. That place is therefore the best for wheat and bees, says Triad 26w.[55] Bronze Age and Neolithic monuments including a stone circle, still stand today on the nearby pig-shaped hill, Mynydd Gwyn; they were probably raised to commemorate her mythic achievement.[56]

The Great Boar, Twrch Trwyth

The Arthurian version of Triad 26, termed 26w, also states that 'it was prophesied that Britain would be the worse for that womb burden',[57] (as if the arrival of Neolithic agriculture was bad news). Consequently,

Sabrina, joined by a Roman image of Ceres on a coin found at Aust Cliff, a prehistoric moon deity from Spain, and the mythic Welsh boar.

CERES

DIVA AUGUSTA

Arthur assembled the army of the island of Britain to seek out and destroy Henwen'.[58] After she had eluded him, he turned to confront one of her possible partners, a gigantic boar named Twrch Trwyth, who, as a king changed into a boar, and came from Ireland with seven piglets. He roamed across South Wales, before being driven into the Severn by Arthur's men near Beachley. According to the *Mabinogion*[59], Arthur ambushed the boar 'with every experienced soldier in this island', and, between Llyn Lleyan and Aber Gwy (the Wye), they closed in on him. They grabbed him first by his feet and soused him in the Hafren, until it flooded over him.[60]

Here we should pause to consider that boar's many riverine connections. There are five Welsh rivers named Twrch, two each in Dyfed and Gwynedd, and one in Powys. The boar was the landscape's running power, while in Irish, *triath*, means 'wave' and 'boar'.[61] Thoroughly soused in the Severn, Twrch Trwyth, the Great Boar, was identified with, and helped to cause the Bore's tidal wave; after serving as Henwen's partner. [62]

When Mabon son of *Mellt*, 'Lightening'[63], took the shears from Twrch Trwyth's head, the boar lost a symbol of kingship. Both the shears and scissors were wrested from him. Yet in the process, Osla Big Knife let slip his knife into the water, which filled his scabbard and it pulled him into the river, where he drowned.[64] This well-known symbol of sexual union underlies the many instances of Iron Age deposition of such weapons into British rivers.[65]

But then Trwrch Trwyth escaped onto dry land at Aust, still wearing his comb, as a feminine emblem, on his forehead bristles. On landing at Aust, he could hardly fail to notice that set into the Triassic sandstone were numerous gigantic 'tusks' of white alabaster, a semi-crystalline form of gypsum, occurring naturally, as if to welcome him ashore.[66] On glancing northwest he could see the 896-feet-high summit of Gray Hill, (Mynydd Llwyd), seven and a half miles away.[67] And at dawn on Lammas, the Welsh *Gwyl Aust* start of harvest, quarter day, on August 5th, sunrise forms an alignment between Aust and Gray Hill. [68]

Then, pursued by Arthur's men from Aust, Twrch Trwyth ran through the modern counties of Somerset and Devon, and the entire length of Cornwall, before eventually heading back into the sea, so completing aspects of Henwen's outward journey, but in the reverse direction.[69] He was truly her mate.

Llyn Lliwan, a Lake of Origins

The lower reaches of the Troggy Brook, up which Henwen had swum when pregnant, featured as in Nennius' *Historia Brittonum*.[70] He explains that when the Bore comes, gathering itself in the estuary prior to running up the Severn, the Troggy receives part of that flood, in the form of a whirlpool, only to disgorge that water in one gigantic mountainous wave, when the tide retreats.[71] Llyn Lliwan is often spelled *Llyw*, a word that has several meanings in Welsh, including: 'female pudenda, fish tail, food, and nourishment',[72] so bringing natural phenomena and animal births together. Likewise in Irish, while *torri*, and *torraf* signify a breaking wave, *torraid* is 'to farrow a litter of pigs'.[73] Likewise, here on Severn-side, the Bore's wave can have positive, birth-giving qualities.

Therefore Henwen's journey up the flooded Troggy could charge her with fundamental life-giving opportunities. Its turbulent waters, contributed by Sabrina, were recognised as an original, comprehensive life pool.

Moreover, the stream bed was composed of cavern-pitted Karst limestone, in which, even today, local people recognise what they call 'Whirly-holes',[74] leading to an underground reservoir, from which a 'Great Spring' gave 23 million gallons a day, creating a notable connection with underworld powers.[75]

After the Severn's railway tunnel had been dug to Sudbrook and opened in 1886, the Great Spring was diverted to feed a paper mill and a brewery. Prior to these disruptions, the Troggy, and Llyn Llyw debouched into the Severn close to the Sudbrook Iron Age camp's ramparts. The Bore-fed Llyn Llyw continues to form at times between Caerwent and Caldicot, in a two mile long, by 300 yard (274 metre) wide expanse, and retains a salmon fishery at Caldicot. So, on both sides of Sabrina's river, mythic narratives attach themselves to the landscape's natural endowment. (Sudbrook Camp is now half eroded by the channel tides. It was occupied by the British, between 200 B.C. and from

A.D. 50, was then used as a Roman supply base, for their invasion of South Wales.)[76]

Yet according to Geoffrey of Monmouth, writing in the 12th Century, Llyn Llyw retained pre-military qualities and preferences. 'If an army faced the lake, and its spray touched their clothes, the entire force would be drawn into it and drowned, By contrast, if an army turned its back on the lake it would do them no harm'.[77] (Perhaps the Romans took note of this when establishing their Troggy-side town of Caerwent). Geoffrey concluded that the Severn Bore *was born* at Llyn Llyw.[78]

The importance of the spot where Troggy entered the Severn is emphasised in the former Welsh habit of measuring their entire country along a line from Anglesey to Portskewett, (considered a major harbour),[79] in fact a village overlooking the Troggy-Severn confluence.

Considering Gray Hill, A Sow in Labour

In a country full of hills and mountains, what made Gray Hill's 273 metre high summit stand out? Welsh *Llwyd*, 'gray', also means 'holy, blessed', but from southwest to northeast, its flat top measures little more than one and a half miles, yet is crowned with prehistoric features. These range in date from a Neolithic 'causewayed camp' enclosure, to Bronze Age standing stones, one of which, over two metres tall, is studded with white quartz pebbles. Of the two stone circles, one contains 12 megaliths, with two more stones, now lying within its 12 metre circumference.

There is also a Bronze Age field system, and a cemetery of five late prehistoric round barrows, some 40 feet in diameter, together with cairns; containing stone cists.[80] Artefacts found here include numerous leaf-shaped arrow heads, axe fragments, and Neolithic potsherds, all providing evidence of Gray Hill as 'an ancient necropolis, venerated for thousands of years'[81]. Part of a large causewayed camp enclosure of a Neolithic type also graces the hilltop. In the 19th Century, hilltop bracken was burned on August 1st each year, and on the Hill's north flank, a Black Sow's Well was probably named in honour of Henwen's frightening winter-time persona,[82] while Brideswell wood, and Brideswell, both on the Hill's west face, bring in the Irish goddess Brigit, arriving to support the divine sow.[83]

Gray Hill's steep-sided isolation offers panoramic views over the Severn-side lowlands, and the river Severn, towards a clearly visible Aust Cliff. From Gray Hill, looking down the estuary, the coastline is visible as far as Exmoor, and even Bodmin Moor in Cornwall. In this manner, Gray Hill presided over an expanding Sabrina, and to river travellers, the Hill cried out to be noticed, not least because the life-giving Troggy curled around its eastern foot.

Salmon to the Rescue

The third of the three prisoners listed in the Welsh Triads is Mabon, son of Modron, 'who was taken when three nights old from his mother. No one knows where he is, or what state he is in, whether dead or alive.'[84]

When King Arthur arrived outside Gloucester, he was addressed from its wall by Eidol, son of Aer, who implied that since Mabon was a child of a goddess, who was as old as the Universe, it was necessary to find and seek help from the oldest living creature, in order to rediscover her son's whereabouts.[85] Thereupon Arthur contacted a man who was renowned for understanding all the languages of wild creatures.

This interpreter duly interviewed several beasts, one after another, collected from different parts of Wales – a blackbird, a stag, and an owl. Each of them in turn boasted of the immense age of their species. Eventually the interpreter interviewed an eagle who recalled how he had sunk his claws into a salmon at Llyn Llyw, and that fish had then dragged him under, till, on the point of drowning, he had released his claws from the oldest and wisest of all creatures. Seek the help of that salmon, advised the eagle, before it flew away.[86] The process, described, in the *Mabinogion*, is prescient of Darwin's achievement, in linking all species together.[87]

Who could be better qualified to find and lead the rescue of Mabon than the oldest and wisest of creatures, leaping from the very pit of creation, at Llyn Llyw? Not least because she travelled annually up the Severn, as part of his normal routine, and in the company of thousands of other salmon who had just migrated across the Atlantic, and were aiming to spawn in the river's headwaters.

But the trip was dangerous. Along both banks stood fishermen with their nets, hoping to pull her out, cook and eat her. The special fish also had to contend with the turbulent waters around the Wye's confluence with the Severn.

At this point, like many an 18th Century traveller, she might become so entranced by the view as to decline to go further. 'Imagination cannot form an idea of anything more beautiful, than what appears here to your ravished sight. The noble river Severn and the boundless prospect beyond, such a bewitching view, that nothing can exceed this amazing reality.'[88]

But if she continued, she would soon risk confronting the temple of the mighty fisher god, Nodens, at Lydney. Nodens considered himself controller of all traffic on the waterway.[89] He also regarded himself as a personification of the Bore, on which he rode, (as depicted by a Lydney image,) crowned with Phoebus-type rays, in a chariot, pulled by four white horses. The horses were accompanied by two winged figures typifying the winds, and two fork-tailed marine monsters – tritons. The salmon avoided Nodens and his many acolytes, lined up to receive cures at his Dwarf's Hill temple, built in A.D. 364.[90]

To bolster her fragile courage, the fish took heart from the Broad Stone, an 8 feet (2.4 metre) high effigy of the Great Goddess, which stood at Stroat, close to the river's North bank. This prehistoric menhir was said to have been used by the 'devil', in a game of quoits played against Jack of Kent, a Herefordshire hero-brigand, who was set into Forest of Dean folklore.[91] (The salmon suspected that their quoit game was an attempt to simulate solar and lunar movements, as seen also on hot cross bun markings. People will do anything to join in the cosmic dance, the fish thought.) But couldn't they see that at the Broad Stone, they were fiddling with an image of Mabon's divine mother?

The fish was also worried by several unfavourable reports that she had heard about the human inhabitants of the forest that flanked the Severn's west bank. Rudder termed them as 'barbarous and emboldened to commit many robberies and outrages, particularly along the banks of the Severn'.[92] Even in the 19th Century, they were 'prone to use charms, incantations, and witchcraft'.

Another practical difficulty involved finding where the river was! Prone to burst its banks, it often formed itself into numerous swamps and lakes, as Leland discovered, on his travels down its course. In addition, major floods in 1483, 1563, and 1606 had caused widespread damage, with the deaths of thousands of sheep, horses, and cattle, together with 'considerable loss of Chrystyian life,' as the Rector of Almonsbury described.[93]

There were also many rival war bands, some from foreign countries, prowling up and down the valley, like those seen by Rhonabwy, who described: 'a brilliant white troop, and a mantel of white brocaded silk about each man of them, and the fringes of each one pure black… Those are the men of Lychlyn (Norway). And a second troop, pure black, fringed with white, who were the men of Denmark'.[94] While appreciating their dress sense, the fish feared their spears.

Yet none of these threats compared in intensity with that which threatened to overwhelm the salmon as she approached Gloucester. When two miles short of his goal, she heard a faint tremor in the water, which soon grew into a threatening roar, from which she realised that he was about to be overtaken, with or without Nodens on top, by the famous Severn Bore. When the Romans first encountered the phenomenon they assumed that the end of the world had come.

The Severn Bore as a Wave of Life

Thanks to the funnel-shape of the Severn estuary, seawater tides regularly surge up the river as far as Gloucester in the form of a great wave or 'Bore', a display enhanced at the equinoxes, in March and September, especially when the full or new Moon's gravitational pull aligns with the Sun's position. At such times, travelling at ten miles per hour, the Bore can be more than six feet high.

In Nennius' 9th Century manuscript, *The Wonders of Britain*, he sees this recurring event as 'the Two Kings of the Severn… two heaped up wave crests …[who] fight each other like rams, one going against the other, (rather than four horses,)… This they do from the beginning of the world to the present day'.[95]

'This admiration of water; it must come from some religion, I suppose', remarked a middle aged lady as she awaited the Bore's arrival at Stone Bench in 1986, while in Rowbottom's words of 1947, it produced 'angry ostrich plumes of white spray'.[96] The Bore, also known as *Hygre*, 'Giant', carries logs of timber and the carcasses of the animals that it has drowned, while 'the spray fills your nostrils with a faint smell of the sea'.[97] The Giant travels up the re-united river as far as Tewkesbury.

Sir Charles Blagden gave an 18th Century account of the Bore's impact on 'the southernmost point of Alney Island, universally called the *Parting of the Water*, where the people of Gloucester assemble, in order to see the tide come in'.[98] Blagden notes that the Bore's arrival is preceded by a roaring sound, audible to those who put their ear two inches above the river's surface. Then the spectators witness 'the dashing of water with great violence over the distant banks', where some men in boats meet the tide and some dogs are thrown in, just as it comes, to observe their howling and distress. Suddenly the boats and dogs are instantly raised up and thrown into violent agitation. At the same time a vast wave or wall of water, reaching across the whole of the Severn and leaping everywhere over its banks, is seen approaching with extreme rapidity. It breaks upon the point of the misleadingly termed Lower Parting (where, in fact the two channels of the river reunite), with a vast surge and a prodigious noise.

This event represents some kind of erotic orgasm, expressed on an environmental scale. The wave, instantly divided by the south, or lower parting, then rushes up one or other of the Severn's two channels. In this way it serves to redefine Maes Mawr, the Great Meadow that has given its name to Maisemore, the Saxon village which overlooks the north end of Alney isle, which is called the Upper Parting. (The island is now named after a Saxon eel fisher, with his eels.) In Celtic myth, 'the Great Plain' was the domain of a Goddess, encountered by everyone posthumously. Therefore, faced with this journey, a dying man hoped to be dressed in his best boots.[99]

In fact, Alney isle, which is a little over two miles long, forms a uniquely large isle in the length of the Severn's 220 mile long course. Accordingly, Alney alias Celtic Maes Mawr, may have been regarded as the surface epiphany of the maternal goddess, Modron.[100] Here she surfaces as a summer-time meadow, now subdivided by hedgerows, while annually retiring, when inundated under her own winter flood waters – even without the Bore's contributions of sea-water.[101] To rise above these, the east-west causeway across Alney from Gloucester's West gate was raised on a series of arches. So the divine mother hovered between solid land and waters of both marine and fresh kinds. She was a mother of the elements, (including her summer-revealed river-deposited earth) which at other times reflected both gleaming sunshine, and the subdued sparkle of stars and moon, while offering a winter playground to salmon and elvers.

From within his stone prison, her son Mabon was closely attached to this elemental array, for as Leland noted, 'The river of Severne brekethe into two armes in the goodly medowes a litle above Glocester, whereof the principal arme strikethe hard by Glocester towne syde',[102] which the imprisoned Mabon overlooked, sited then as now above Gloucester's riverside quay. But his prison walls prevented him from seeing or appreciating his good fortune. Yet with Alney Isle creating his mother's solid, water-fringed outline, he could hardly have been placed closer to her side. Modron as Alney Isle is the mother goddess's anciently acknowledged, midstream epiphany, adjacent to her son. Together Mabon and Modron made a river-scape version of the classic Christian Mother and child image. But perhaps he was howling because he was too close to her side, and wanting instead to move away, in order to enjoy independence, with the approach of springtime.

Releasing Mabon, Son of the Goddess Modron

After many hours spent negotiating the Severn's capricious meanders and sand banks, and having survived all that the Bore had thrown, while the weight of the two knights that she was carrying on her slippery shoulders grew tiresomely heavy, the oldest and wisest of creatures finally pulled up alongside Gloucester's Roman quay, where, sure enough, the piteous cries of the gaoled Mabon rent the air.[103] After a perfunctory survey of those prison walls, the two knights lost their enthusiasm for the task, and decided that reinforcements would be necessary, before any attempt could be made to free the infant.

They ignored the salmon's protest that delay was unthinkable. She realised that Mabon, in his youth, *had to participate* in the young year. And as every salmon and eel-mother knows, the stream of the year always sets the pace. Once missed, you were stuck. Consequently King Arthur's table was round, *circular,* designed to keep in tune with world and cosmic events, followed by every living thing. For that reason, one of these two knights was named Cei, and often given the epithets *gwyn* 'fair' and *hir* 'tall' – fair and tall as a decorated Maypole. Some derive his name from *cei* 'quay', perhaps the very Gloucester 'quay' to which his destiny had carried him![104] Moreover, his companion Bedwyr was closely linked to Welsh *bedw,* 'birch tree',[105] which in Wales was regarded and treated as a leafy Maypole.

But the two knights refused to attack the gaol. Instead they remounted the fish, and were carried back to Arthur to report their findings. Arthur promptly 'summoned the warriors of this island of Britain', to attack Caer Lowy, which they did.[106] Then Cei, perhaps animated by shame after his earlier reluctance, tore through the prison wall, put Mabon onto his back, fought off the prison guards and escaped. So Arthur came home, and Mabon with him, a free man.[107]

And of the three prisoners in the Severn listed in the Triads – namely Gwair in Lundy; Llyfr, underwater near Aust; and Mabon in Gloucester; only Mabon is released.

At Gloucester, the participation of the oldest and wisest of creatures had lent a guarantee of primal authenticity to that event. Mabon represents the world's seasonal release from the dark 'underworld' of winter. He is therefore the annual temporary re-emergence of the Lundy Isle's imagined permanent abundance. In Mabon, an Other World ideal coalesces with a joy filled surface world reality. Touchingly, Mabon is protected by the Roman Army's veterans, installed at Glevum from its foundation, where each man was given a smallholding to cultivate, in a return to farming after their many battles. Thus they had a direct engagement with the physical fabric of the island-wide goddess, Modron.

When society, and indeed reality as a whole, was mythically grounded, the basic elements of Earth, Air, Water and Fire were embodied, and dramatically brought together within and as supernatural mythic characters. What, then, among this cast of interacting beings, did Mabon contribute?

Mabon as Maponos-Apollo, a Sun God

The Celtic god Maponos was worshipped in Northern England and Southern Scotland. His name stems from Gaulish and Old Welsh *mapos*, 'young boy or male child', plus the 'on' particle, signifying 'divinity'.[108] Together they make Maponos, *the* supernatural 'Great Youth'. The 6th Century *Ravenna Cosmography*, compiled from earlier sources, lists *Loca Maponi*, 'The Place of the Youth-God', as a centre for his worship.[109] This is feasibly identified as Loch*maben*, a small loch-side town in Dumfriesshire. (Its name has drifted from *Mapon*, via maben, towards Brythonic *Mabon*.)

Eighteen miles southeast from Lochmaben, and overlooking the Solway Firth shore near Gretna Green, stands the mighty Clach*maben*stane (Mabon's stone) – a mineral embodiment of the divine Maponos. This ice-born granite megalith, more than 7 feet (2.1 metres) high, and 18 feet (5.5 metres) in girth, is one of only two surviving stones in a former prehistoric circle of nine uprights, remembered by mid-19th Century people.

In 1982 the Maben Stone itself toppled from its shallow foundation pit. Oak fragments contained therein gave archaeologists the chance to obtain a radiocarbon date of c. 2525 B.C. for the circle, before they re-erected the monolith. So it seems that the original 'Mabon' is another child of the Neolithic or New Stone Age Goddess,[110] known throughout Eurasia, and represented architecturally at the Avebury and Silbury monuments in Wiltshire.[111]

Clochmabenstane is positioned at the north end of the 'Inlet of the pillar ford', an ancient route across the Solway, used by the Scots and English, who continued to negotiate at the sacred stone throughout medieval times. As with Mabon-on-Severn, Maponos also stands on strongly tidal waters. Therefore it is no surprise to find his name, *Deo Mapono*, clearly incised on a moon-shaped silver crescent, found at Vindolanda, on Hadrian's Wall.[112] When the British (proto-Welsh) were driven out of the area c. A.D. 650 they carried their devotion to Maponos-alias Mabon with them into Wales.[113]

Prior to that, the Romans had equated Maponos with Apollo, the Greek god of light. In northern Britannia they set up altars, four of which survive, engraved with joint dedications to this merged pair, viewed as a single deity.[114] One such altar, from Corbridge, describes Apollo-Maponos as *Citharoedus,* 'The Harper'[115], in recognition of their shared musical and poetic skills – gifts that continue to play an important part in Welsh life.

From this heritage, it is likely that Mabon at Roman Glevum (a town given a name by its Roman founders in the 1st Century A.D., which signified 'Bright' or 'Shining').[116] Mabon-Apollo was installed there as a youthful manifestation of the sun god, desperate to be released from imprisonment from night and winter's darkness, in order to shed his beams upon the extended body of his local water mother, the Maes Mawr of the river Severn deity, which he overlooked.

Though Alney is now crossed by railway lines and pylon-high electric power cables, its place in the affections of Gloucester people, as expressed by the town's poet, Ivor Gurney, remains in high esteem.[117] Alney, in its entirety, displayed the water-and-earth body of the Goddess.

BAG LADY – MOTHER NATURE IN HUMAN FORM

A seated female figure made from green plastic rubbish bags, stuffed with mid-October's fallen lime tree leaves. She is a compost heap of one year's decay, which contributes towards the vigorous growth of next year's vegetation, including human foodstuffs. Similarly, the parabolic garden canes that support her portray the rise and fall of the unified Life-Death cycle.

Bag Lady is Mother Nature in human form, sitting on a 'birthing stool', habitually used during parturition by many cultures. The umbrella depicts the firmament of a billion galaxies, her starry offspring.

If seen as the Virgin Mary, her Son, born on Earth, may become

Bag Lady, after an autumnal image by Michael Dames (2008).

a plastic football-as-cosmos, since Christ-the-God includes all imaginable matter, rolled into one ball. This sculpture combines both Christian and pre-Christian notions of sacred reality together with the mundane equipment of modern secularism, in an attempt to discover an overall synthesis.

BRIDE'S SCOTTISH HOME

In this essay I claim that in the Scottish Isles, the inhabitants literally dwelt inside their cosmic outlook, by designing their houses to match the female shape of their Universe creator's body.

'One feature of Scottish mythology is the predominance of goddesses. They are greater and stronger than the gods', according to D.A. Mackenzie.[1] They are written into the forces of Nature, and include tempest, thunderstorm and raging ocean, where their presence is confirmed by the Gaelic language. The environment manifests *the Cailleachan Mor*, or Great Hags, whose presence was also discernible within womankind. The Landscape Hags matched Scotland's rugged terrain, and so were often portrayed as 'crude and cruel fearsome monsters, or as bloodthirsty avengers, whose anger was appeased by offerings and magic ceremonies'.[2]

Until recent times, formal human behaviour was modelled upon the presumed acts of the deities, who were themselves embodiments of recurring natural events. Consequently humanity as a whole was immersed in the divine epiphany of Nature and in a cosmos regarded as sacred.

The interplay between mountain hags and womankind is implied by the Gaelic *Bean/Beann* words. Thus *bean*, 'Woman, wife, she-goat, active, nimble', lies alongside *beann*, with the meanings 'skirt, drinking cup, mountain peak, headland, and pinnacle'. Similarly Gaelic *Mam*, (derived from Gaelic *Mamaide*, a child's word for 'Mother'), is also 'a handful of meal' and 'a large *breast*-shaped hill' *and* 'a pass between two hills', (comparable to the Earth Mother's vagina).[3]

Of the many tales concerning mountain building hags, one tells of the hag who set off for Knock Farrel, with a great creel of earth on her back. When the creel bottom gave way, the resulting cascade of earth formed the Little Wyvis Mountain. The *Cailleachan Mor* are inseparable from the physical fabric of Scotland.[4]

As Alexander Carmichael makes plain in his book *Carmina Gadelica*, physical and spiritual realities merged as one for the population of the Scottish Isles. There, 'religion, whether pagan or Christian, permeated everything, not least their homes. The people were sympathetic and synthetic, unable to see, and careless to know, where the secular began and the religious ended, being endowed with profound feeling in their religious instincts'. He found that Highland divinities are 'full of life and action, local colour and individuality'.[5]

Of these deities, none registered more pervasively than the Goddess Brigit, alias Bride, whose worship had been introduced from Ireland. To the Gaelic speaking Scots her name meant 'The Exalted One'. She was regarded as the pioneer of manifold skills;

an innovator in metalwork, weaving, dairying, medicine, and the domestic hearth. She was also revered as the patroness of poetry, while nurturing and protecting cattle and sheep,[6] especially when on the highland summer pastures. Bride also presided over holy wells, often named Brideswell in her honour. Powers of divination and prophecy were attributed to her.[7]

As in Ireland, she and her two sisters formed a triple goddess group, in which she took the Virgin's part, and so, after the arrival of Christianity, she was often coupled with the Virgin Mary. Yet Bride remained fully involved in practical affairs. In her there was no separation between inner and outer realities; instead, the mundane and the spiritual were completely merged. Due to her influence, domestic and agrarian life were woven together.[8]

Accordingly, with each reviving year, Bride's energy was seen as a serpentine life-force. On the eve of her February 1ˢᵗ festival, pitched midway between winter solstice and spring equinox, 'early on Bride's morn, the serpent shall come from the knoll', that is to say, from the supernatural world of the *síd*. 'On the day of Bride of the white hills, the noble queen will come from the knoll to offer a hope of winter's end, and the flow of new milk from livestock'.[9]

On the eve of February 1ˢᵗ, every household prepared for her arrival, by providing a specially prepared crib, in which the goddess would be laid, in the shape of a doll, dressed in white, ceremonially carried into the house by the head of the family. Thus, as a snaky life essence, the goddess assumed infant human shape. In doing so, the deity had again crossed the reptile-human, dark-light boundaries to initiate another spring-time.[10] To accommodate this event, which was staged in innumerable houses, each house had typically been shaped, when built, to match the outline and features of Bride's fast-maturing adult body. Consequently, the family could literally dwell within their goddess. From the door, where a human bride was greeted on her first entry with a cake, broken over her head, to the central hearth, and the often stone-corbelled 'head' chamber, (simulating the divine skull), human and divine lives were lived as one.[11]

By occupying the body of their belief, the householders hoped to share in her benefits of a two-way relationship. For example, when the fire was dampened down for the night every evening, and, (regarded as an extension of Bride's own sacred flame), it was never allowed to go out, the housewife sang: 'I will smoor the hearth as Brigit the foster-mother would smoor. May the foster-mother's holy name be on the hearth, be on the herd, be on the household all.'[12] When a child was born to the family, 'the baby was handed to and fro, across the fire, three times, and then carried three times sun-wise, around the hearth', so bringing sun, hearth-fire and infant together. It was also claimed that Bride had nursed the infant Christ, 'the Noble king of glory on her breast'.[13]

With her many fire-laden associations, Bride's shape was depicted in many small vessels, including the steatite oil lamp found on Farleyer Moor, near Aberfeldy, and in similarly shaped lamps from Kettlestone and Horkstrow brochs.[14]

Her presence was also embodied in the tools concerned with weaving and corn-grinding. Thus the third verse of a Hebridean croon was addressed to the household's rotary quern: 'Bless the gear and bless the handle, the tackle, each part that belongs to it.' The same chant continues: 'Put sun-wise turn ever on quern, if thou do wish white meal ever to be plentiful. Turn her in showers with sweet liltings'. Such songs conveyed a unified outlook, suffusing everything.

As their recorded prayers show, Bride was also their doctor or aid-woman. 'Aid thou me in my unbearing. Aid thou me, great is my sickness'.[15]

In Kennedy-Fraser's words, 'The civilisation of the hearth is a social order symbolised by the beautiful figure of Bride. It employs images and emotions which carry the memory back thousands of years.'[16] And her 'thousands of years' is no exaggeration. According to the archaeologist, C. Gordon-Childe, who excavated the Neolithic Orkney village of Skara Brae, the culture of the late 19th Century Hebrides closely resembled that which he found in the 3rd millennium B.C. There too, the acts of a goddess, discerned in Nature, were *visualised* as images, in sizes varying between figurine and temple ground plan.[17]

Moreover, the maternal house-image is discernible in the overall shape of prehistoric and later dwellings that have been excavated at a number of sites, namely at Buckquoy, Yoxie, Udal, Chlann Eilean

Mor, Hower, Stanydale and The Gairdie, in addition to the no. 8 building, at Skara Brae, which Childe interpreted as 'accommodating that communitie's ceremonial activities'.[18] Individually, and as a group, spread across many periods and races, these structures offer a substantial body of evidence, supporting a superhuman anthropomorphic intention, underlying house design. This is matched and supported by the small-scale female images, of clay or stone, found in some of these dwellings,[19] and by the harvest-time corn dollies, made from the 'last sheaf' of the year, and still regarded as *The Caïlleach*-in-person, that continue to hang in house or barn.[20]

At the Links of Noltland settlement, Orkney, three such solid figurines were found in 2012, of which one was located within a specially built household shrine, a small, clay-lined stone recess.[21]

In Shetland, the archaeologist Charles Calder made a convincing comparison between the largest Neolithic structures that he found on those islands and the Goddess-shaped temples of Neolithic Malta. On Shetland, he found variants of the 'large lady' ground plan used in houses, temple, *and* in chambered tombs (overleaf).[22] Bride's image apparently offered a 'cradle-to-grave' reassurance and protection, given the harsh climatic conditions often prevailing outside. Yet, as may be inferred from the pattern of field boundaries drawn onto the back of one Noltland figurine, her benign influence extended into the farmland, and was just as concerned with the prosperity of agriculture, herds and flocks, as the numerous 19th Century appeals made to Bride on such matters, reiterate.

Yet some of today's experts are reluctant to accept the Noltland idols as divine, since that would smack of *idolatry*. The idols have been trivialised as 'only toys' or 'just a bit of fun'.[23] Is this an example of modern archaeology putting its faith in today's mechanistic science-ism and imposing its own culture onto the past, rather than acknowledging the essence of what has been lost? As Mircea Eliade, a great expounder of mythic realities, sadly concluded when he said 'the ability to transcend our own time... has been completely rooted out'.[24]

It is hard to account for archaeology's recent ignorance of Scotland's rich and well-documented folklore. Whatever the cause, the outcome is a repeatedly wasted opportunity. About 100,000 ordinary people

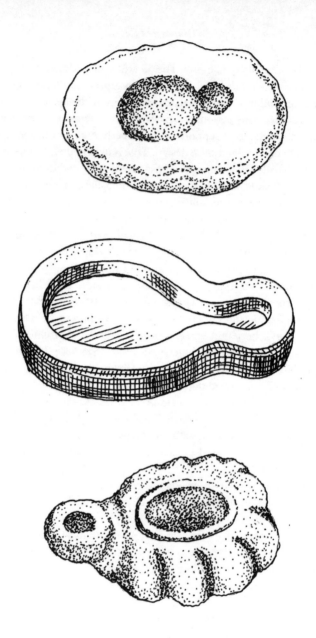

Mother-shaped stone lamps (top) from Okstrow, and the Tealing House, Scotland, along with a corn-grinder (bottom) from Finting, similarly shaped.

queued up to catch a glimpse of the Noltland goddess images when they were briefly put on public display in 2008, only to be fobbed off with official evasion, as to the statuettes' true significance.

A similar nervousness was shown by the archaeologist, Anna Ritchie, during her 1976 excavation of house no. 4, at Buckquoy, which she classified as of 'figure of eight' type, while admitting that 'the term is not ideal'.[25] Another expert has redefined Scotland's humanoid shaped houses (a preferred shape that persisted well into the Bronze Age), as 'Circular disc, plus rectangle, plus square indentations'; another exercise in denial, in which geometry replaces Ritchie's 'figure of 8'. The unwillingness to register an organic, functioning unity seems to be their shared priority.

Yet in 'considering the possible anthropomorphic qualities of the Buckquoy house', Ritchie went so far as to invite comments from an anthropologist, Anthony Jackson. He assured her that 'all over the world, the structure of houses often reflect the occupant's religious views'. He cited the Dogon people of West Africa, 'whose homes mirror their cosmology'.[26]

But for a strictly scientific investigator, the validity of the 'results', depends upon maintaining a scientific division between a 'subject' reduced to 'data', and a detached investigator, rather than a living member of the human race. Ritchie's chosen methodology was almost bound to produce the exact opposite of the occupied house-as-deity *experience*.

One way to by-pass this impediment to earlier norms of understanding has been provided by Charles Thomas. In his study of Iron Age and Pictish art motifs found in Scotland, he points to the frequently employed image of a stylised human figure, holding aloft a whirling solar disc.[27] This is a theme that features in both Scandinavia and Scotland, and provides a miniature template for the sun-related Bride, whose house 'head chamber' may provide a solar reference.

Kennedy Fraser compares the voice of Celtic Paganism to that of the deep sea, which endlessly circulates beneath the brief tides of Christianity and modern instrumentalism, just as the hypnotic croon of their melodies seem to have no ending, but create a rhythmic pulsation, inviting self-surrender to the curve of their tunes.[28] Listen to the words and the music sung when Bride was invoked on May 1st, as the herds and flocks were driven to the upland summer pastures. 'Come, fair Mary from the cloud, May Bride the calm keep the sheep. And propitiate to me the cow of my love.' Another prayer is addressed to 'Bride, smooth white of the ringlet locks,' before the reaping starts, and again prior to hunting. While yet another chant begins; 'from the holy maiden Bride, radiant flame of gold'.[29]

Bride was in turn maiden, mother and hag. As Mother, Gaelic *màthair*, she is also the 'source' and 'primary cause', while as *math*, she is 'good, happy, glad'. Is it no more than a coincidence, that the Gaelic word *brigh*, 'substance, essence, juice, sap', and Gaelic *brig* is 'to build'. Brighid, Brigit, Bride, perhaps it is not too late to welcome you home again.

Writing in the shadow of the World War I catastrophe, the poet Laura Riding concluded that 'the word Woman' describes how things are *as a whole*, and the final unity. She referred to the 'inside sense' delivered by the 'world within houses, a female space... in which men and women may jointly move and live'.[30] Therein, the disruptive male emphasis on difference, particularity and history, can be absorbed into the feminine regard for universality, imbued with a rhythmic consistency, and the achievement of repose. If that is true, the value of Bride's house-building project becomes more than 'a thing of the past'.[31]

THE GREAT BRITISH COASTAL WALK

The fluidity of Britain's place names, and their interrelatedness with the landscape, can provide us with a useful insight into our colourful past. The Celtic languages in particular have made their description of British geography into a narrative.

To walk around the intricate coast of Britain involves a journey of up to 7,000 miles, according to some estimates; it involves passing through several languages, including pre-Celtic tongues such as

Pictish, (overlaid by Gaelic), Welsh, and Cornish, Scottish, Anglo-Saxon, and Norse. The beaches are cluttered with the flotsam of these cultures, which deposit a shared anthropomorphic attitude towards coastal and inland features, revealing their belief in a humanoid, living topography, as recorded in place-names.

Place-names imagine coastal rock stacks as horses, or as cows and calves, pigs and piglets. Also featured are pre-Christian and Christian references, including the names of saintly hermits' refuges, intermingled with those addressing the Devil. This debris of beliefs, displayed in words and landforms along the strands, sits alongside names linked to modern commercialism, and military defences, in a tide-washed melee, wet and dry, checked by thousands of groynes and breakwaters, trying to reduce the longshore drift of shingle, under the pull of the moon.

To start at the end, we could go to the Land's End peninsular in Cornwall. There one finds *The Horse,* and nearby a *Horse Rock,* and a *Tobban Horse,* perhaps three reminders of the Celtic sun god in equine mode, plunging into the western sea at dusk. *Horse Rocks* are a feature of our coastline. There are dozens of them; plus several *Mares* and *Colts* and *Grey Mares.*

As well as many horse names we have the so-called White Horses, the white crested waves that crash against their stone 'fetlocks' at every high tide, when the ocean contributes to the apparent animation of such creatures, helping to define the horse as the essence of power and speed.

Credited with the ability to propagate, a *Bull Rock* often rises close to a *Cow Stone,* sometimes named *Blind Cow* or *Bo Cow.* These 'Cows' in turn, can have a smaller *Calf Stone,* as in an alignment off Swyre Head, Dorset. There the Blind Cow is also credited with producing the nearby Butter Lump rock stack. For generations of pastoralist farmers, these sea-born 'cattle' *worked.* They served as the marine prototypes of all living cattle. At least 150 million years old, such effigies were partly reborn from their oceanic womb, at every low tide, and were probably revered as icons.

In addition to such bovine images, *Sheep Rock,* and *Goat Rock, Little Ox, Sow Stone* and *Piglets, Dog* and *Bulldog,* also feature around our shores.

So the pastoralist's livestock spilled over the cliffs and, like sacrificial offerings made to the sea, stand permanently on the beaches.

As a collection, this entire group of salt-sprayed animals have represented farmers' livestock, for 6,000 years, through the length and breadth of this island. They are dynamic, noisy, threshold beings; huge liminal creatures, positioned to welcome the first (and successive) waves of Neolithic agriculturalists from the Continent, into the virgin farmlands of Britain, from at least 4000 B.C. onwards.

They seem to give Nature's permission for the inland agricultural performance to start, based on their own elemental combination of sun-splashed, wind combed, rock and water.

Coastal place names make it clear that these species must also share the land with wild animals, such as the frequently named *Wolf* and *Bear*, along with *Rat, Bat, Hare, Buck,* and *Otter* – all ossified into coastal stacks.

As for *men*, near Redruth there is a *Deadman's Cove*, (one of many), a reminder of the multitude of shipwrecks and loss of life that has occurred around our perilous shores. If *The Man and his Man* islands also hint at a forgotten tragedy, so too might *The Old Dane*, who has *St Agnes Head* for company, and *St Piran's* ruined church and monastery in the vicinity, ready to pray for his immortal soul. Further west is *Hell's Mouth* bay, opening as if to swallow the nearby *Ceres Rock*, commemorating the female Roman harvest deity of that name. *Peter's Point* and *The Three Brothers* stand guard over that life and death encounter.

Cornwall also offers several examples of the primal creation Goddess in a fish-human merger. The Mermaid readily crosses the sea-land boundary, and in doing so, she sings alluringly about the unity of all life forms, as implied in her own half-fish body; and she will plunge to any depth to prove her point. At Zennor, in West Cornwall, her image is carved onto a 15th Century bench end in the village church. From her home at Pendower Cove, she had sat on that bench, while a chorister, Matthew Trewella, fell in love with her. They ran off together, and he was never seen again. He willingly drowned in her oceanic Everything-ness. However reluctantly, we must all repeat his disappearance in the goddess's waters of renewal.

To round England off as a royal domain, *Crown Isle* lies close to

Cape Cornwall, which used to be regarded as Britain's most Westerly point. Cape Cornwall overlooks a remarkable Early Bronze Age barrow, on the cliff edge at Ballowell. Its elaborate chamber was probably viewed as an entrance to the Other World, associated with the Scilly Isles. Formerly termed the Silly or Blessed Isles, the Scillies lie in the Atlantic Ocean, 20 miles beyond Land's End. As in Ireland and Wales, such offshore westerly isles were typically seen as supernatural realms of eternal youth, (Irish *Tír na Nóg*), presided over by a benevolent 'Lady'.

Consequently, on the Scilly Isles you can find *Great Arthur* named, along with *Arthur's Head.* Was he sent there to recuperate, after his death in the Battle of Camlan? During 1970s summers he was joined by Harold Wilson, then British Prime Minister, seeking refreshment on *St Mary's,* the biggest isle in the archipelago. From there, they could both enjoy *The Great Cheese Rock, The Tea Ledge, The North* and *South Cuckoo, The Great* and *Little Smiths, The Moon Rock, Droopy Nose Point, Barrel of Butter, Lion Rock, Brewer, Baker,* and *Nut Rocks,* plus *The Mare, Cow* and *Calf, Doctors Keys, Maiden Bower, Queen's Ledge,* and *Golden Ball.*

Facing them, close to Land's End, rises *The Irish Lady,* a wave lashed pinnacle, to which an Irish woman, the sole survivor of a shipwreck, briefly clung; but she was washed away before rescuers could reach her. Now, as a rock, she stands there indefinitely.

Flanking the very tip of the same peninsula stands *Dr's Head,* close to *Dr Johnson's Head,* each with their own story to tell. They are guarded by *The Armed Knight* reef. As the author of the first English Dictionary, *Dr Johnson's* headland, so positioned, implies that the entire island of Britain, and indeed every material thing is, at root, only a tenuous verbal construct; yet *firmly* rooted when, as here, Dr Syntax is at hand, to articulate this country's grammar and sentence structure. In every down-to-earth matter, language has the last word, because things change, according to how they are defined. The land begins and ends with speech. This is what educated people have come to imagine, while Dr Syntax sends a warning to those beachcombers who find significance in the merest fragments of old rubbish.

In any case, much of the nomenclature along the Cornish coast and Scilly Isles is meaningless to an English traveller, since it is inscribed in

Cornish, a Celtic language that died out in the 18th Century. Instances from Scilly include: *Iniswilig, Minalto, Biggal, Minmanueth, Muncoy, Menbean, Wingletang, Lethegus, Isinvranc, Crebawethan, Annet, Creeb, Gannink, Mouls, Hanjague* and *Pernagie,* with many of these names recorded in medieval manuscripts. How eagerly do the salt waters, lying beyond history, snatch at the chance to return our firm but transitory definitions to non-meaning!

Conversely we are equally prone to haul in the riches of the ocean towards land, and to attach them to coastal features. So we have *Crab, Lobster, Mackerel, Bream, Limpet, Bass* and *Tortoiseshell Rocks, Sharks Fin, Minnows Islands, Pilchard Cove, Conger* and *Cockle Stones,* plus *The Dolphins,* and remains of some *Ancient Fish garths,* and *Fish Markets* listed, along with *Mussel* and *Oyster beds* and *Pits.*

Meanwhile, a perpetual gathering of a million sea birds, keen to share in this feast, is recognised by the terms *Mew Stone* and *Great Mew Stone.* There they sit and call; to be answered from many a rock named after *Gull, Crane* and *Heron,* or from *Goose* and *Shag Stones.* So the British coasts advertise a plenitude of interdependent life forms, including humanity, seen and heard enthusiastically devouring one another, at a perpetual meal.

To ascertain whether this gobbling and exchange of life between species, has any *meaning,* our coastal place names attempt to attach the entire island to the Christian faith, and thus to a belief in a transcendental holy spirit, that hovers over this island-world of gross matter. As a result, many of our most prominent headlands are called after devoted followers of that religion, and offer to sailors and mariners a solid topographical 'rosary', a string of implanted prayers, to send them safely on their way.

For example, *St Alban's Head,* Dorset, celebrates Britain's first man martyred for his Christian faith, c. 500 A.D., while *St Catherine's Point,* Isle of Wight, combines her chapel with a modern lighthouse, in a techno-theological amalgam that reflects her own martyrdom on a flaming wheel. *St Edmund's Point,* Norfolk, holds the remains of his chapel, while *St Abb's Head,* Berwickshire, carries a Nunnery. *St Bee's Head,* Cumberland, *St David's Head, Ann's Head,* and *St Govan's Head,* (all in Pembrokeshire), bring the west coast traveller back via *St Thomas'*

Head, in Somerset, and from there, to the afore mentioned *St Agnes Head* in Cornwall.

This array of divine intermediaries has probably been much depleted in recent centuries, in keeping with the population's gradual drift away from Christian commitment, led by the post-Reformation Church of England. Thus *St Mary* and *St Margaret* have now to be content with only a *Bay* each, in Sussex and Kent respectively. They are submerged, while the former *St Mary Rocks* (both West and East), in Plymouth Sound, have now lost their saintly prefixes. Meanwhile, *St Anthony* still makes his *Point* in Cornwall, and has two Rocks elsewhere. *St George* (of course) has his *Isle*, and *St Lucas's Leap* is a natural arch in Dorset – altogether a somewhat meagre collection for the *Old Priest* and *Clerk Rocks* to supervise.

By contrast, the earliest phase of Christian influence, represented by offshore island hermitages, still features strongly on our maps. From *St Clement's Isle* and *St Michael's Mount*, Cornwall, to *Lindisfarne*, and the *Farne Islands*, Northumberland, and *Iona*, in Argyllshire, the locations of these semi-detached hermits' cells present a curious paradox; for in seeking to escape the 'world', the solitary monk often retreated to where the elements conduct their daily drama, in the most inescapably vivid manner. Were the solitary pioneers, in going to the extremes, worshipping a transcendental God, or Mother Nature at her most compelling, or both-as-one? In raising the issue, the monks left their unanswerable questions. Whether or not God still lives in, or under, *Maison Dieu Hill,* near Southwold, Suffolk, (perhaps along with fairies, termed 'The Mother's Blessings' in Wales), is equally uncertain.

As for the work of The Devil, He appears on the coasts as *Devil's Arch* and *Devil's Bridge* in Wigtownshire; *The Devil's Threshing Floor,* Kirkcudbrightshire; *Devil's Bank,* in the Mersey; and as *Devil's Hole* and *Devil's Kitchen,* near Clovelly, Devon. Perhaps in Britain's long existence, the Christian interlude represented a very brief and exceptional dream of somewhere else, namely Heaven, and its opposite, Hell. Other aspects of this tendency to seek somewhere else may be seen in the coastal names *Newfoundland,* Somerset; *Nova Scotia*, Northumberland; *California Gap*, and *Gibraltar Point*, the last two in Norfolk; with all four hoping to sail off to sunnier opportunities, by superimposing an alternative destination onto the here and now.

Those of us remaining here are left to wonder, who is the *Meg* that sits on *Meg Rocks* or *Meg Stone* at several points around the water's edge? Why are *The Dancing Beggars*, playing in the sea, off Dartford, Devon? Will *Old Harry* ever reappear at his Dorset Rocks? Can *Jupiter* be expected to live up to his name, in Plymouth harbour? Has the Phoebe of *Phoebe's Point*, Devon, departed for ever, leaving only her first name behind? How did *Tilley Whim* get his Dorset *Cave*? And do the *Seven Sisters*, the range of chalk cliffs near Beachy Head, Sussex, exchange greetings with the *Three Sisters* of Midlothian, or with *The Sisters* of Newquay, Cornwall? Are they all related, as mermaids turned to stone? Did *Old Parker* lose his *Cap,* near Dover, in saluting them?

Was *Shakespeare's Cliff* in Dover named after the spot to which Gloucester believed he was led in Shakespeare's tragedy *King Lear*? Is *Ladies Hole Pain*, near Sittingbourne, Kent, incurable? Is there nothing more to say of *Jenny Bells Carr*, Northumberland? For what did *St Margaret Hope* in Fife? Will *Maggie Brady* please return to her *Rock*, near St Andrews, and ask *The Mermaid* to sit again on her designated *Chair*, alongside *Lot's Wife* in Kirkcudbrightshire? Is *Meg Merrilees* in *her* nearby cave, close to *Daft Ann's Steps*, in Dumfriesshire? What does *Meg* do with her Berwickshire *Dub?* And does the Berwickshire *Maiden's Stone* belong only to her? Will *Stinking Scar,* in Cumberland, stink evermore? And did *Crazy Mary* go mad because she was shut into her own *Hole* near Lowestoft, Suffolk? Is *Damehole Point* near Bude, Devon, still occupied? If so, is she on speaking terms with the nearby *Beeny Sisters*? And why are so many anonymous *Ladies* identified with low tide ledges and hidden coves around our shores?

But instead of trying to answer these questions, perhaps we might instead take pleasure in the total obscurity of these faint traces of superhuman or subhuman narratives. Britain, it seems, is an untidy ragged country, especially around its edges. However, I am inclined to continue to gnaw on the rocky bones of these mysteries, and hope to enjoy some of their marrow. As a scavenger, I must feed on meaning, wherever it lies, not least within recent displays of commercial and military power.

The British coasts are scored all around, with the named evidence of mercantile, military, and holidaymaker's energies. The docks in London, Liverpool, Glasgow, Bristol, Hull, Southampton, Middlesbrough, Felixstowe, Leith and Aberdeen are terminals of a recently vanished world Empire, spanning one third of the globe, and with some ports heavily involved in the Africa to American slave trade. All these routes were flanked around the coasts by dozens of shoreline and offshore lighthouses, coastguards, lookout posts, and lifeboat stations. A stone on the shore, east of Brighton, commemorates the setting up of the world-web of longitude lines, based on 0 degrees, running from the Thames-side Greenwich Observatory.

In addition, the island was defended from invaders by coastal forts, ranging in age from Roman, through the Tudor, and Napoleonic era, to those erected during World War II. They are most densely clustered around the southeast corner of Britain.

From there, stretching towards East Anglia, the Romans built a string of 4th Century forts, to protect the 'Saxon Shore' from us, the invading English. Our first landing is commemorated at Pegwell Bay, Kent, in A.D. 449. Henry VIII's forts, constructed from Dover to the Solent, were aimed at all-comers, and fired at the Spanish in 1589, the Dutch in the mid-17th Century, and the French, especially from 1790 till 1815, when a series of so called Martello towers were added to the array.

To prevent attacks upon Britain's naval bases at Plymouth, Portsmouth, and the Firths of Forth and Clyde, more forts were built during the 19th and 20th Centuries, sometimes on offshore islets, to fend off France, Germany and the U.S.S.R. And all these mighty efforts are now largely obsolete, since the next invader is likely to arrive in the shape of a rocket-delivered hydrogen bomb.

Meanwhile, numerous pipelines bring crude oil from the North and Irish Seas' deposits, (now somewhat depleted,) to several refineries and storage depots along the coasts, while 'wind farm', clusters are erected offshore.

After the health-seeking example of the 18th Century Hanoverian British monarchs, and subsequently with the help of new railways, the people of Britain flocked to her newly built coastal resorts for their summer holidays. There they sat huddled in flimsy glass and iron shelters, watching the rains fall, for a week or two. Yet however inconvenient this collision of the elements proved, it refreshed the lives of these visitors who had arrived burdened with urban toil.

However, after the 1960s with the arrival of cheap air transport, the population tended to flee to the Mediterranean for their vacations, and most of the British coastal towns acquired a semi-redundant tranquillity, unsure when (or if) their show would ever start again, while their cast iron Victorian piers, built for entertainers, collapsed, one by one, into the sea.

What this brief survey suggests is that every era, decade, and day, is a lost cause. While the boom of surf keeps the walker company, it sounds the drum-beat of perpetual change. However much it *looks* like the ever-renewed lace of Britannia's petticoat, it usually *acts* as the bared teeth of a ravenous marine lion, forever spitting out bits of its previous land meal.

Place names reflect this ceaseless process of removal and replenishment. Yet each human generation is given a fresh chance to briefly experience *a lasting* attachment to their island home, by sharing in, and by *combining with* its physical fabric. The clear invitation is embedded in so many place names, both coastal and inland, making a union with these land's forms almost inevitable.

When a promontory projects from the main mass of the land into the sea, just as the human head rises from the main mass of the body, it is perceived and named as a head or headland. There are at least 19 examples in Cornwall alone.

Flamborough Head and *Spurn Head* carry the same habit north into Scotland where *Kirktown Head, Scotstown Head, Rattray Head, Kinnards Head* and *Troupe Head* address the North Sea, while *Ness Head, Duncansby Head* and *Dunnet Head,* surround and convert the northern most corner of our island, into a triple headed monster.

This type of identification is so entrenched as to pass almost unnoticed. It lies deeper than metaphor, in a zone of instinctive affinity,

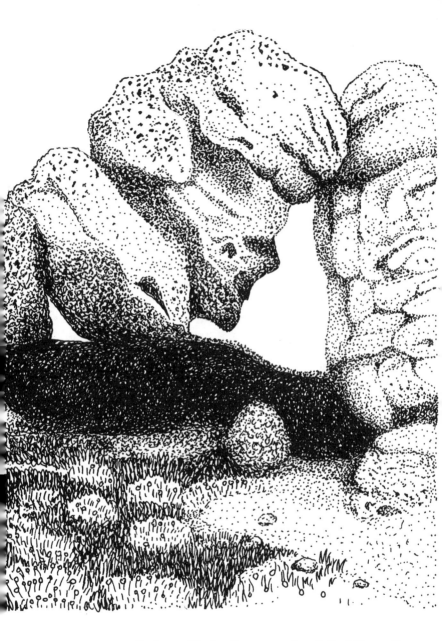

Dr Syntax Head, at Sennen, near Land's End, Cornwall.

usually taken for granted. It offers both a means of re-attachment to the island, plus, by extension, a standing invitation to *belong* to the entire material universe, providing a solid link to everything beyond the self.

In this reunion, there is no such thing as *dead* ground. And indeed, without the organic life of the headland's living soil, together with that same thin layer stretching inland, we would physically starve and collectively soon die out, even if each atom of underlying rock is fizzing with enough energy, (as is now known), to match *The Dancing Beggars* beyond Dartmouth.

In this Great and vital Britain, the headland can also register as a stabbing, 'pointing' finger, hence *Poldui Point, Lizard Point, Manacle Point* and many others. With *Lower Sharpnose Point* and *Upper Sharpnose Point*, Devon, finger and nose combine, for the Earth can arrange its features in flexible combinations. Though at Torquay's *Hope's Nose,* at *Long Nose Spit* in Kent, and *The Nostrils,* Isle of Wight, the projecting nose may also feature alone.

Perhaps *The Seven Sisters,* who show off their bare white torsos as the magnificent chalk cliffs near Eastbourne, before running inland as a series of modestly grass-covered ridges and valleys, are tacitly inviting the entire country to share in their animated state.

The notion of a 'living land' features in both Old English (O.E.) and Old Norse (O.N.) contributions to modern English.

After centuries of estrangement, by subdividing analysis, this inheritance is still there, urging us to re-identify with the place-name reminders of land spirits that haunted (and still haunt) many a hill and stream. Thus O.E. *Puca,* and *Pucel* (Puck) are present in place names, along with his alias, Robin Goodfellow (who occupies many coastal Robin Rocks); *Scratta,* a hermaphrodite Devil; *scucca,* a 'goblin'; and *nicor* the 'water sprite'; along with *draca,* the 'dragon 'or 'serpent' of *Worm's Head,* the Gower and *Great Orme's Head,* North Wales.

O.E. *hoh,* a projecting 'heel of land', as shown by 'a giant, lying face down,' is incorporated into Ivinghoe Beacon, Hertfordshire.

Scotland has its own collection of these emanations, including Water Bulls, and Water-horses, ever-ready to interchange between animal and human shapes. The *bean-nighe,* or washer at the ford, prepares a shroud for a person who is soon to die. *Tacharan* is a diminutive male water-sprite, while the *Kelpie* is also fond of river-banks, and manifests as a beautiful young woman. By contrast, the *Beithir* who can appear as a serpent, or as Lightning, lives in mountain caves and corries. The Gaelic words *frid* or *fried* are applied to both gnomes and elves, believed to dwell in the innermost parts of the Earth.

In the Hebrides gnomes sometimes crowd into a house as densely as a swarm of gnats. Generally well disposed towards people, they appreciate crumbs and drops of milk falling from the table. The *caoineag* (The Weeper) foretells and weeps for those who are about to die in battle. The *Peallaidh* (Brownie) and the *Uraisg* are found in glens, corries, and mountain streams. The *Uraisg* is a friendly half man, half-goat in shape, with long claws. A spirit who supervises all aspects of household weaving and spinning is the *Loireag*, for whom offerings of milk are given. She insists on the women singing in tune while they work.

Also fond of music are the *Cow Maidens*, who tend the sea-cows that are said to wade ashore through the surf of the Outer Isles. Scotland also has many tales of *selkies* – seals that are able to shed their skins to adopt human form, and then to mate with, and have children by a mortal. The husband is said to hide the seal's skin, to prevent his wife from returning to the ocean.

These minor Caledonian spirits join the huge number of goddess-related figures in Scottish mythology. Collectively, they are The Hags or *Cailleachan Mor*, the individualised forces of rough Scottish Nature, often displayed in storm, mist and rain. The Hags are celebrated and feared seasonally around the coasts; for example, a hag known ironically as *Gentle Annie*, off Cromarty, was identified with a wind, perilous to fishermen. She blew inland at the end of winter for six weeks, keeping them ashore, until finally relenting to allow the new summer, her opposite number, to take over. To the Celts the natural world was entirely numinous, with each hillock, stream and the sea, credited with psychic intelligence; to them, all running water was sacred.

As a coastal feature, *The Cailleach* becomes a rock stack, close to Cape Wrath, Scotland's northwest tip. There she is paired with *The Old Man*, her consort, as a neighbouring stack. On Colonsay, the hag is

Cailleach Uragaig, (Gaelic 'renewing'). As for her watery associations, there are eight Scottish streams called *Allt-na-Cailleach*, while her name is also attached to a *Creag,* (a cliff); a *Cnoc,* (a hill); a *Meall,* (a round hill); a *Beinn,* (a mountain); a *Coire,* (a corrie); a *Lochan,* (small loch); a *Gleann,* (a glen); a *Strath,* (a broad valley), a *Sron* (nose, promontory) and a *Tom* (a round hillock). In the Highlands of Scotland, the Goddess can ride anywhere.

A spiritual feeling pervades throughout ancient descriptions of watery landscapes. This attitude merges with anthropomorphic coastal references, while inland the same life-death legacy provides the norm. Thus *torr,* 'a conical hill', holds both extremes of *torrach,* 'pregnant', and *torradh,* 'burial'.

The word '*Pap*' (breast) is often applied to pointed hills, as in the *Meikle* and *Little Pap,* overlooking Glen Muick, the *Pap* overlooking Loch Lomond, and the well-known *Paps* of Jura.

Indeed a glance at the human-landform interplay, revealed in the Gaelic and Welsh languages of Britain, indicates that an immanent yet supernatural quality was installed throughout the land, long before the arrival of Anglo-Saxon and Norse peoples added to the same tradition.

For example, the Welsh word *bron* ('breast,' and 'seat of thoughts and emotions'), is also used to describe no fewer than 85 breast-shaped hills in Wales, while the word *pen,* both in Welsh and Cornish, is both 'Head' and 'Headland', plus the 'sources of both the intellect and of rivers,' in 2,100 named instances. Welsh *moel* (bald), is used in Mynydd Moel, (a bald, rounded, skull shaped mountain), while *moel* additionally serves as 'a heap of corn', or 'butter rising above the rim of a vessel'. In the same language, *Braich,* 'arm' is used 25 times to describe either 'a ridge of land', 'an arm of the sea'. (*Braich* can also be 'a line of verse', where thing and thought combine in harmony.)

Forty named instances of *troed,* 'foot', are found at the base of Welsh mountains. Fully alert, the Welsh word for 'eye', *lygad,* describes, 'the source of many a river or stream', and also appears as 'sunshine between showers', while doubling also as 'a viewpoint' and a 'medium of expression'.

Wales also has many animated coastal features, such as *March*

Mawr and *March Bach*, 'Big and Little Horse', plus Mare, and other peripheral rocks named after an oxen, a bull, pig, sow, fox, foal, roebuck, roe deer and kid; also a hedgehog, a shrew, and three mice, along with various birds, including raven, kite, gull, and puffin; plus a bishop, a vicar, a witch, and an idol, all topped by high white clouds, that the Welsh refer to as 'the flocks of the (supernatural) shepherdess, *Gwenhudwy*.'

This sample of Welsh place names indicates the innate capacity for synthesis, typical of all the Celtic languages of Britain. They bind the island's coasts, along with the entire land *and* sky, into a solid union with the human inhabitants, both in body and spirit. By doing so, these names reveal a primeval norm of consciousness, reiterated by later waves of settlers, to offer a way of living and understanding that is still available.

Yet all this vitality comes at a price, since what is said to be living, must inevitably also die, perhaps lost at sea. To pick up any handful of pebbles on any English beach is equivalent to entering a Geological museum, for each stone may have been worn from a different rock formation, standing somewhere around the coast, before being transported by longshore drift to this particular spot. Two years ago a friend sent me a postcard from Norfolk, to say that many of the coastal footpaths that she had previously walked on, had, during a single stormy winter, fallen into the sea.

It seems that Britain, as an entity, is a passing fiction, a short play in Nature's endless board-game.

Despite, or perhaps because of, this speed of change, the population has concocted a variety of narratives, in order to find a sense of security here. These fictions typically display a self-centred outlook, yet are often advertised as patterns devised to promote the general good. But, whatever the strategy employed, you may be sure of two things; that any beliefs are just passing through; and any marks you may leave behind here are soon likely to prove indecipherable.

The British coastline as a whole might well be regarded as an example of the free play of non-hierarchical, non-stable meanings, as outlined by Derrida's theory of Deconstruction, in which he goes on to show how, through deconstructive reversal or inversion, marginalised terms can become central. The muddled fringe *is* the centre, and the central truth.

SILBURY'S MISSING MYTHS

How can we claim to be studying Silbury, if we ignore the mythic basis of prehistoric and pre-'Enlightenment' life, that prevailed world-wide, prior to the late 18th Century?

Cosmogony, the birth of the world and of the Universe, is myth's chief preoccupation, in which concern for human birth and pragmatic concerns such as food play only a small part. Most myths invite us to stretch our outlook and boundaries, in order to set pressing local concerns into a universal context.

The principle question posed by myth when operating in a pre-modern society is: what story does that society wish to uphold as being centrally valid to its entire existence? And since every pre-modern society assumes the presence of a divine foundation to its surface events, myth's task is to articulate in narrative form, the relations between divine and ordinary realms.

In doing so, the verbally expressed mythic story is partly inspired by the sacred connections perceived in the place where the story is told, and is normally then supported by visual symbols (including those laid out on an architectural scale), and by the ritual acts and ceremonies that, during seasonally timed festivals, bring these symbols to communal life.[1]

Prior to the invention of written script, the ability to 'read' sacred images, varying in size from figurines to those on an architectural scale, was correspondingly keener, as was the ability to appreciate local topographical features, such as the Swallowhead springs, that could helpfully ground and complement the mythic account.[2]

Around the world, mythic 'truth' is formed from a compound of word, visual symbol, and landscape suggestion, all effectively confirmed by seasonal farming patterns, and that of the human Birth-Maturity-Death cycle, with both types of event being seen as earthly aspects of an underlying sacred reality, involving solar and lunar cycles.[3]

Therefore it should be possible to reconstruct any myth from the surviving strands of such evidence, if they are drawn together by an interdisciplinary search, including material supplied by oral folk tradition, echoing long after the original spoken words have died away.

Yet few British prehistorians, are prepared even to attempt this necessary task of interdisciplinary reconstruction, such are the rigidity of current 'disciplinary' boundaries. Add to that, the science-only methodology that now prevails among British archaeologists, which effectively cuts them off from the myth-based world which they claim to be exploring.

Their approach demands a stance of objective detachment, entirely at odds with the emphasis on participation, which engagement with myth requires. In addition, a prevailing cult of 'rational' secularism discourages them from uttering the word 'sacred', since for them, the supernatural is 'out of bounds.' Instead, as Atkinson boasted before tunnelling into Silbury, they are usually content to measure things, defined in quantities. Consequently, as he declared, after his 1968 excavation into Silbury: 'When the prehistorian comes to matters of faith and religion, he is usually inclined to take refuge altogether in silence.'[4]

Given this situation, the archaeologists Jim Leary and David Field, who led the 2005 repair of Silbury, (following the damage caused in part by Atkinson's futile tunnelling), are to be congratulated for tentatively challenging this sterility. They are starting to consider Silbury in terms that might match the monument's mythic Neolithic origin, among the deciduous forests that largely covered the British Isles, after the retreat of the last Ice Age, c. 10,000 B.C. Field writes: 'We can only hope to understand Silbury if we combine our scientific beliefs (a faith still firmly in pole position) with a vision that goes beyond them.' So, for example, he now suggests that 'the Silbury ditch may have had a metaphysical function'.[5]

The advent of Neolithic farming involved deliberate forest clearance, across the British Isles, an effort credited to goddesses such as Tailtiu. She is said to have toiled so hard on her task that she died from the effort involved in creating a fertile agrarian plain, centred on Co. Meath.

Her dying wish was that an annual festival (Gaelic *Oenach*) should be held in her honour, on the hill Tailtiu, (Teltown), named after her. This gathering was to occur for a week, or at least for three days around the start of harvest quarter-day, on August 1st. And as the annalists report, *The Lughnasa Oenach* began in 3300 B.C., nearly a millennium before Silbury was built. Annual First Fruits harvest gatherings were subsequently held there, even into modern times.[6]

The Teltown deity appears to span the change from pre-agrarian to Neolithic eras. She brings with her a typically Neolithic belief in a feminine power, underlying that revolution. This female orientated veneration persisted throughout thousands of Neolithic years, as female figurines, found across Europe and Asia, suggest.[7]

Tailtiu, one may say, is an embodiment of the land on which she laboured. Her name is derived from a Brythonic (British Celtic) loan word, *teledїw*, meaning 'well-formed, beautiful'.[8] Alternatively, as some authorities believe, her name comes from O. Celtic *talantiu*, 'the great one of the Earth'.[9] Tailtiu is buried at Teltown.

The earliest recorded title of the August 1st gatherings was Ir. *Brón Trogain*, meaning 'when the earth sorrows in pain, in giving birth to its fruits', a metaphor based on a human mother's birth-giving experience, here applied to the month of August.[10] This usage matches the pre-modern belief that the entire landscape, including its rivers, constituted an organic, living entity. The topographic *it* was a divine *Her*, regarded as the goddess of sovereignty.[11]

Just such an attitude is reflected in her gigantic water-drawn effigy that surrounds Silbury Hill. It depicts her in an advanced state of pregnancy, and about to painfully deliver yet another harvest, and another corn idol (modern 'dolly') harvest child.[12]

This is not to claim that the Silbury monument has only one possible interpretation, deriving from its clearly defined Neolithic Great Goddess context. Rather, in standing its ground so well, the earth and water Silbury monument finds itself extending its time-range, in both pre-Neolithic and post-Neolithic directions. Silbury is a world centre that can adapt to change, and absorb its pre-Neolithic setting. So, for example, the *purely elemental* spring water from Swallowhead was carried to, and drunk on, Silbury Hill's summit, as recently as the middle of the 19th Century A.D.[13] Silbury intended to incorporate the pre-human world, and in doing so, she resolves the alleged Nature-Culture divide.

In a similar pre-Neolithic spirit, a poem written in 1007 A.D. specifies that the annual ritual meal served on Teltown Hill consisted entirely of *pre*-agrarian wild foods, including fish from the river Boyne, venison, fruits from the Isle of Man, bilberries from Bri Leith, Co. Longford, and watercress from Brosnach. In this 'wild' feast, the Old Stone Age was saluted, while bread was not mentioned. Instead, the first sheaf of the new wheat crop was ritually buried at the site, in recognition of Tailtiu's work, in preparing the ground for its eventual cultivation. Thus it, and nearly 100 other 'Harvest Hills' acknowledged Nature's gifts, both wild *and* cultivated, with bilberry picking still a feature of these assemblies into the 1940s.[14]

Forty-two wild plants or 'weeds', were found growing on Silbury in July 1857 by Professor J. Buckley.[15] A survey carried out a century later by 18 Wiltshire botanists, more than doubled this number of plant finds. They found an additional 43 wild plant species thriving on the Hill.[16]

The botanists also recorded 10 species of grasses growing on Silbury, including two kinds of 'wild oats' grass, and a proto-barley grass, named 'meadow barley'.

All these 'wild' self-set species, which reappear annually on Silbury's flanks, originated long before humanity arrived on Earth, (and of course they predate the entire Neolithic era by millions of years). So these annually self-renewing plants, which are mostly in flower at Lammas, confirm the Silbury builders' plan to see the world as more than an exclusively *Human* concern.

Yet as local terms for these plants make clear, the rural population of Britain saw our native wild plants offering innumerable cross references with the animal and human inhabitants of a shared planet. In this 'poetry of interconnections', wild plants invite us into a unified living world.

The 95 self-planted vegetable species found on Silbury in 1957 included those with folk names including: Horse hard head, Pony tails, Ox's eye, tongue and lip, Cow Parsnip, Pignut and Pig Mouth, Hog Weed, Sow Thistle, Sheep's Bit, Lamb's Tongue, Goose Grass, Chickweed, Dogwood, and Dog Rose, Cats Tails, Harebell, Birdseye, and Bird's-Foot-Trefoil, Hawkweed, and Hawk's Beard, Crow Garlic, Toadflax, Mouse Ear, and Adder's Tongue – benefiting from the Sun and Moon Daisy, alternative names for the Ox-eye flower.

While such names build bridges between Silbury's plants and the animal kingdom, they also display many links to human life. For instance, we find Lady's fingers, in no less than five species, and several instances of Lady's slippers, while Lady's Bedstraw features twice. In addition, there are particular references to Queen Ann and to Creeping Jenny. Maiden Heads and Virgins also crop up, along with Sweethearts, Mothers, and Mother's milk, plus Baby, and Sucky Sue. Old Woman's pin cushion and Dead Man's fingers also feature. As regards the social classes, Silbury's plants provide a range running from King and Queen, via Lords and Ladies, to Chimney Sweep.

In addition to the Barley and Oats grasses, other farming issues are depicted in plants, found on Silbury, named Rest Harrow, Hay Rattle and Haycocks, while our breakfast is delivered by the Hill, through species known as Bread and Cheese, and Eggs and Bacon. As for curative properties, many of Silbury's plants were believed to serve as self-healers, wound healers, Eye-brighteners, or as a cure for constipation.[17]

Post-Neolithic waves of immigrants into the British Isles brought with them a switch of emphasis towards male deities, such as Lugh, who eventually renamed the start of harvest gathering after himself, as *Lughnasa.* Lugh was the bright young god, with an underworld rival named *Crom Dubh,* 'the crooked dark one', bent under the weight of carrying ripe sheaves. He came to represent the fertile underworld in which the crops had grown and were attached, until Lugh sharpened and swung his sickle, to cut off Crom's corn heads.

The Lughnasa Oenach

The Lughnasa Oenach at Teltown was Ireland's most famous August 1st gathering, and the symbol of a nation-wide High Kingship. Seen from Uisneach, 28 miles (45 km) away, the Lughnasa sun appears to rise over Teltown and the same primary beam also passes over the Sleave Beagh mountain, named after Ireland's mythic founder, Bith, meaning 'Cosmos'. This sunbeam reaches the sea at Dunany, named after the goddess Aine. Until the 1930s, horses were submerged in the Blackwater river, before dawn on August 1st.[18]

It is said that Lugh had placed an effigy of Crom's head on top of Teltown Hill, before he began work. This stone head represented the corn, topping all the ripe straws harvested by Lugh and his human followers. Yet although Crom Dubh lost his head during this process, the 17-feet (5-metre)-high, 280-feet (85-metre)-diameter, platform on the summit of Teltown hill, called Rath Dubh, is probably named after him. That enclosure resembles the flat top of Silbury Hill.[19]

Yet from her hilltop grave, Tailtiu continued to lend a peaceful tone to the proceedings. A truce was maintained during the gathering; disputes and fights were banned, as were the theft of yoked oxen, and the slaughter of milk cattle. In place of endemic violence, hopes of corn, milk and good weather were offered. The mood combined expectation with joy when it was fulfilled. The ceremonies included music, wrestling, hurling matches, and 'entertainments for mind and body': also horse races, and horse submergences, in the river Blackwater which looped around the site. As long as this *prime-oenach* continued, 'Ireland would not be without perfect song', the medieval poet concluded.[20]

Crom's crop growing power was derived, he believed, from Maebh, the goddess of Connaught, whom he acknowledged as his queen. In England, his presence was recognised in Cornwall by Eleanor Hull.[21] While Crom and Lugh's joint annual task was taken on, and Lugh's name features in the Roman name of Carlisle, and as *Lleu* in the Welsh *Mabinogion*, across England, Lugh becomes the John Barleycorn, who, since Tudor times, has featured in our folksong. And this Barleycorn derives from the Anglo-Saxon god, *Beowa*, meaning 'barley'.

At Silbury, these male deities functioned on and around its Neolithic Goddess effigy. Similarly, the female *Cailleach* figure, an embodiment of the deposed Neolithic Goddess, continues to haunt the Irish *Lughnasadh* sites, as at Slieve Gullion, Co. Armagh, where she has a rock chair, and her house was believed lodged into that hill's summit.[22]

In post-Neolithic periods, and in keeping with the rise of patriarchy, 'the presiding king became the great figure at Teltown', say medieval reports, suggesting either the provincial king of Connaught, or the high king of all Ireland – the latter being often a somewhat insecure post. Silbury's equivalent 'royal maleness', is found in the shape of King Sel, Sil, or Seall, recognised there as an embodiment of harvest. In 1668, Samuel Pepys gave a local man a shilling for this information.[23] King Sil's appearance fits a male-female, king-earth goddess union, often seen as a *hieros gamos,* a sacred marriage between the eternal and supernatural earth queen of the territory, accepting a temporary mortal monarch as her consort.[24]

Whatever Silbury's original outlook, she now incorporates the masculine into her cultural heritage. From the Iron Age onwards, kingship, patriarchy, and some male gods, have altered the surface of her goddess-centred world, while her moat body and her harvest hill-womb have survived these changes. So she stands for all our prehistories, and serves as an omphalos, one of the world's central points, uniting time and space.

At Teltown, such was the fundamental necessity of their annual harvest First Fruits festival, recognised as Tailtiu's gift, that when St Patrick arrived on the scene in the 6th Century A.D., he was unable to ban it. Instead, he established a church at nearby Donaghpatrick, and thrust Laoghaire, the presiding king, into a 'dark condemned hole' near the river.

Likewise at Silbury, three millennia after the Hill was built, some plant names were changed to accommodate Christian influence, and now incorporate Adam and Eve, Mary, and the Devil. Similarly, due to Christian pressure, the August 1st festival was renamed Lammas, 'loaf mass', when the first grains of the new harvest were baked and blessed, in celebration of Christ's Last Supper. Further, by the 19th Century, the date of the Silbury gathering had been switched to coincide with Christian Palm Sunday in Easter Week, so marking and repeating Christ's arrival in Jerusalem.

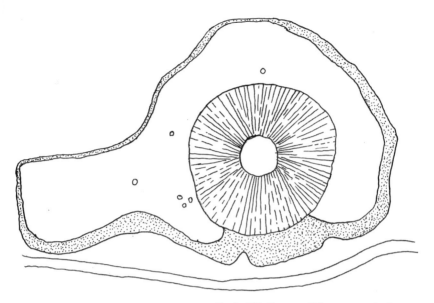

*Early 20ᵗʰ Century Silbury moat shafts,
which proved the Neolithic date of the monument.*

On that day, a large number of temporary side stalls lined the main road below Silbury, reported the oldest surviving Avebury inhabitant, who was interviewed in the early 20ᵗʰ Century, and quoted from his written memoire. He claimed in 1900, 6,000 people met at Silbury Hill for dancing and other sports.

The stalls apparently sold toys, sweets, nuts, ginger beer, flat round gingerbread cakes, and much prized Lent Figs. Hundreds of people, dressed in their best clothes, came from all the surrounding villages to this festival, where there was much merry-making and beer drinking.[25]

The principal amusement consisted of a rude kind of tobogganing, in which the young lads and lasses dragged planks of wood to the top of Silbury Hill, and on these slid down to the bottom, much to the detriment of their nether garments.[26] In this manner they dropped into the same dry section of the ditch between its causeways where John Barleycorn's First Fruits had previously been laid. So the Hill's ability to deliver children, both male and female, was continued, though surrounded by cornfields far from ripe at that April date. Thus myth's flexibility ensured Silbury some kind of a long life, with Christ, coming on Palm Sunday, to be sacrificed and then resurrected, thereby extending the role of John Barleycorn, mourned and celebrated in the still well-known British folk-song:

> There was three kings came into the East
> Swore that John Barleycorn should die.
> They took a plough and ploughed him down,
> Put clods upon his head.
> But the cheerfu' Spring came kindly on.
> And shours began to fall;
> So John Barleycorn got up again
> And sore surprised them all.
> The sultry suns of summer came
> And he grew thick and strong
> His head weel armed with pointed spears
> That no-one should him wrong.
>
> ...His colour sicken'd more and more
> He faded into age
> [His enemies then] took a weapon sharp
> And cut him by the knee
>
> ...They wasted, oer a scorching flame'
> The marrow of his bones
> But the miller used him worst of all
> And crushed him between two stones
> And they have taen his very heart's blood
> And drunk it round and round
> ...Till their joy did more abound.
> ...Twill make a man forget his woe;
> Twill heighten all his joy.[27]

The interweaving of death and resurrection, displayed in human and vegetable life cycles, is also displayed in Silbury's physical fabric. She was considered as a heap of rock containing the accumulated remains of countless shell creatures, that lived and died beneath the Cretaceous ocean, 60 million years previously. Through them, we are returned to the remorseless fact of mortality. Consequently, Silbury can appear as an all-time funerary monument, where empty shells support the barely discernible traces of many human generations, united as one. Yet the Hill is nevertheless offered to the living, by symbolically containing the unborn child of the next generation, together with the ripening crop, necessary for its, and all human sustenance. This new food was made available as First Fruits every August 1st. Both the organic plant and animal remains, discovered stored inside the Hill's core, combined with the rich flora of its surface, together suggest an irrepressible life-giving capacity, resulting from the productive struggle between surface world and underworld, including its deathly aspect. This linkage has been depicted in works of art, such as Richard Long's length of Silbury string, which when coiled up on a dark floor may depict the unborn child of the next generation, and, when extended, turn into its maternal umbilical cord, set against eternal blackness.[28]

Regardless of the disappearance of the Goddess from the minds of most British archaeologists, her moat-image outline persists around the Hill, as neatly as every mother carries her unborn child. Although there has been a big fall in the level of the local water table since Neolithic times,[29] Hill plus moat combined to make the monument, continue to provide us with what Nietzsche termed 'The eternal return of the same' and 'eternal recurrence'.[30] That 'same' is the world as a whole, seen in magnificent, intensified miniature, to offer both the pilgrim and the casual wanderer a supernatural account of reunified life on Earth, set at home within the cosmos.

With this comprehensiveness, Silbury prefigures Blake's New Jerusalem. On Silbury, that writer's 'Albion', a male embodiment of Britannia, shares this birth mountain with Jesus Christ, and with the corn dolly named King Sel, symbol of the annual harvest. He, in turn, is coupled with the clownish John Barleycorn of folk-song fame, who provides *and personifies* the malted barley's ale and the effect of its consumption.

The arrival of Christ during Silbury's Palm Sunday festivals ensured that *his* Bread and Wine, literally transformed during every celebration of the Catholic Mass into his flesh and blood, continued the Neolithic tradition of divine figures, as superhuman manifestations of essential food and drink, grown on and around the Hill. Christ's reported descent into Hell on Easter Saturday brings the underworld into his story, and eliminated Crom Dubh's role. Christ's subsequent ascent into Heaven likewise rendered Lugh's function obsolete, at least for Christians. Yet when Samuel Pepys visited Silbury in 1688, after more than 1,000 years of Christianity, he was told by a local man that the monument was named after King Sel, Sil, or Sael. These names are derived from Old Norse *saell*, 'happy', and Old Teutonic *Saeli*, meaning 'Good, or Blessed'. Both words are incorporated into Middle English, as 'happiness, prosperity, good Fortune', and 'a favourable time for action', as in 'Hay Sele', and 'Wheat Sele' harvest times.

Thus the rural population met the advent of Christianity, by naming their monument after a secular king, who took over the harvest-time role of the prehistoric gods, *and* their human servants. King Sil, buried within Silbury, performed therein as another 'Crom Dubh', yet he also reappeared on the surface at harvest time, to become the divine Lugh's reincarnation. He therefore doubly helped to sustain Silbury's mythic continuity throughout the Christian era. All these characters were born from the maternal earth, wherein harvest chaff mixed with Mary's afterbirth debris, were returned to the ground, and contributed towards the fertility of the soil beneath our feet, so producing an overall tolerant norm.

Beyond the tragedy of sacrifice, animal and vegetable ingredients contribute to a *pan-theo-mime*, or 'dance of the gods', with deities cavorting upon the only available stage, provided by our well-watered ground, as James Joyce described. As one of the elders among this great company, ever-maternal Silbury can continue to provide her well-timed fecund moments, gathered into timelessness. So the lost idea of The Great Goddess provides a map for our futures. We cannot do without our earthly foundation, or push it away, into any plausible concept of *The Past*. That being the case, I say: 'Welcome again, you Stone Age deity. I still think the world of you, and wish to be properly earthed'.

Silbury confronts us with the assimilation of opposites, whereby the seeming incompatibility of cyclical and linear time schemes is resolved. On that Hill, the mortal king's linear reign also maintained and epitomised the annual cycle. Likewise, Lugh's sunlight, flashing between the equinoxes across the body of the Silbury moat, is tied to his dive below the horizon every night, there to encounter subterranean powers, including bent Crom Dubh, while the future king must embrace a repulsive underworld hag, as a prior condition of sovereignty being conferred on him.[31] She then confirms her choice with the cup of water that she offers him. Her act was echoed by a ceremony, held yearly on Silbury as recently as the 19th Century, when Swallowhead spring water was consumed on the Hill's summit. Moreover, as 17th Century folklore, as reported by John Aubrey, declares, 'Silbury was built while a posset of milk was seething'. Suggesting that Silbury's new king was to be seen as the mother goddess' child.

Silbury emphasised a night and day union, with her First Fruits arriving by moonlight, during a *Gwyl Aust,* a night vigil, (hence that Welsh name for the event). Thereafter, as surely as winter follows summer, the goddess-shaped West Kennet Long Barrow cast its mid-day shadows in Silbury's direction. So Silbury encapsulates a dynamic stability, reflected in the behaviour of the stars.

In this task Silbury is assisted by each of the neighbouring Avebury monuments. As a group they describe a *cyclical progress* in tune with the annual passage of the seasons.[32] And their old plan is both discernible and available, ready to help defuse our recent craving for unrestrained linear progress, and its dangerous consequences.

The combination of earth, water and sunshine and moonlight, brought together on the wind-ruffled surface of Silbury's moat, is now often concealed by the 16-feet (4.9-metre) deep layer of white alluvial clay that now clogs the moat, as A.C. Pass found in 1886, when he sank ten shafts into its surface, to reach the chalk bedrock.[33] Yet despite this material, introduced over more than a 4,000 year period by the moat's inflowing waters, today the water level often still rises above these layers of silt. (Pass found Neolithic worked flints among the silt's lowest layers, with Bronze Age material at a higher level. In addition, in 1968, Atkinson discovered 100 high-value Roman coins in the moat.)[34] The ditch still floods during spells of wet weather.

During a recent three year-long survey of Silbury's exterior, David Field paid scant attention to the moat, yet in 2003 he was prepared to write that 'Silbury's ditch was more than a mere quarry', and that 'it was intentionally filled with water, to produce a mirror-like surface, inviting shamanistic ceremonies of a metaphysical nature'.[35]

Leary later added that these 'water-related' ceremonies may have occurred around what he termed 'the cistern', (alias the neck and head of the Silbury water goddess image), and concludes that 'the Silbury ditch may have been as important as the mound'.[36]

Coincidentally, Steve Marshall, a musician and writer who lives near Avebury, noted that the same thin layer of clay rich material (that geologists term Plenus Marl), was feeding water into the Silbury moat. The same clay causes the nearby Swallowhead, Waden Hill, and Pan springs, to spurt from the chalky underworld. Marshall noted that these waters always emerged at 50 degrees F (10 degrees Celsius). This was also the temperature of the 82 jets of water that he observed 'bubbling beautifully' around the eastern edge of the Silbury moat in January 2003. After Marshall invited Leary and Field to witness his discovery, Field found seven more such outpourings on Silbury's eastern flank.[37]

The mystery of Silbury's valley bottom siting has surely now been solved. It was determined by the need to fill with water the reclining Goddess effigy, defined by the moat, enabling her to become pregnant, as demonstrated by the hill that she surrounds. That Hill was raised almost entirely from material drawn from her own body. So the deity known from innumerable Neolithic pregnant figurines achieved monumental form.[38] And on her, the entire community's hopes of prosperity were believed to depend.

After pondering the value of the moat image, Leary now recognises that 'the inverted reflection of the hill would seem to disappear, as through a mirror, into the Underworld,' thus completing a cycle between realms. He might also have mentioned that her water filled body can hold moving reflections of clouds, sun, moon, and of distant stars, thereby drawing the entire cosmos into her body. Being aligned due east-west, her moat also maintains a perfect balance between equinoctial sunrise and set, giving equal 12-hour long shares of day and night, darkness and light.

Nor did Silbury as a deity neglect the practical need for water. Rather, her functionalism was comprehensive. It involved quenching the thirsts of people and their livestock, and watering their crops, while providing river water for transportation and fishing.

Given his new-found interest in water, Leary has recently 'wandered down to the Swallowhead spring'. It emerges barely 500 yards (457 metres) south of Silbury. There he belatedly finds 'that you are entering contemporary sacred ground. The great willow tree, hung with rags, close to the spring, now forms an arch, creating a sort of portal that one has to go through, (if one so chooses), which helps to give the impression that you are entering a different realm.' And indeed he is. There he encounters ordinary people's instinctive reverence for the natural world, a response shared by the prehistoric community that built Silbury.

Exploring further, Leary finds 'crystals, candles, wind chimes, and other votive offerings left around the springhead', signalling the recovery of an officially ignored tradition.[39] Meanwhile, a few academics have been trying for years, with little success, to interest archaeologists in the 'lost realm' of the sacred.[40] Determined to catch up fast, Leary, with another reductionist's headline, has declared that 'Silbury is a monumentalised springhead.'[41]

Meanwhile the devotional trinkets left around Swallowhead suggest that the 'past' is far from over. Our sense of human unity on Earth is confirmed by sharing spring water, and by walking on the chalk downs, in our ancestors' footsteps, while enjoying the same sun, moon, and seasons, and the sight of farmers at work among their crops and livestock, as in the Neolithic. Wild deer continue to run from the nearby Savernake forest, just as they did in the Old Stone Age, long before the Avebury monuments were imagined and built. The Neolithic outlook and achievement lies in the Here and Now, much to the enrichment of our present tense.

VIII.
The Christian Overlay

DEVIL TALK

During the last two millennia, the Christian Devil has been invoked to nullify the Neolithic Great Goddess. She was the central figure of that era, the life giver and receiver of the dead.[1] But the incoming culture has inverted previous beliefs, and she is now linked to the ultimate evil, personified by the Christian-imagined fallen angel, named Lucifer,[2] alias Satan, Beelzebub, the Devil, old Nick (Scandinavian *Hold Nickar*, a sea god, who demanded an annual sacrifice), his Satanic Majesty, Old Scratch (Icelandic *scratti*, 'devil'), Old Harry (English 'to harry', lay waste), and The Deuce (Gaulish *ducius*, 'a demon').[3] Armed with all these titles the devil was invited to arise from his underworld dwelling to poison the land of Britain in its entirety, and to claim the physical remnants of an affirmative Stone Age culture as a province of his negative domain.

Thus he was seen carrying a huge lode of earth and rock in his apron, (note the womanly garment), through Avebury parish, in order to bury the town of Marlborough beneath this debris. He paused, to ask a cobbler how far off his target lay, and the cobbler showed Satan all the worn out shoes in his pack, and claimed that they had all been ruined in his attempt to get to Marlborough; whereupon Satan threw down his lode immediately, and so made what we now call Silbury Hill.[4] This story disregards the fact that the Silbury monument predates Satan's arrival in Britain by more than two millennia, yet achieves the aim of redefining one of the greatest and most positive of Neolithic achievements, the womb of the pregnant Great Goddess, as a heap of misplaced rubbish.

Today, The Devil continues to influence our archaeologists. They now see the Silbury moat, designed as the water body of the pregnant Goddess, as a means 'to keep evil spirits at bay'.[5] And in truth, Satan remains mysteriously pervasive in our largely post-Christian secular world. Thus the long barrow chamber, standing two miles east of Silbury, is called The Devil's Den, while another Neolithic long barrow near Warminster is known as The Devil's Bed and Bolster.

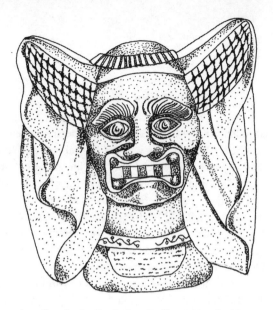

A medieval misericord, from Ledbury, Herefordshire.

A row of five Bronze Age bell barrows in West Sussex are collectively called The Devil's Jumps. In Wales, the capstones of barrow chambers are often referred to as the Devil's Quoit, though elsewhere he delivers a cruel type of play. For example, at Stanton Drew, when called upon to provide the music for a wedding party, he made them dance faster and faster, all night, and then turned bride, groom, and all the guests to stone. Consequently that stone circle is now known as The Devil's Dance.[6]

The Devil's antipathy to the Goddess is evident at the new bridge, Kirkby Stephen, Westmorland. There he can be heard grinding mustard seed, while the female named Jingling Annas, (who reputedly has her hands reduced to stumps by the manacles that she is forced to wear), was also imprisoned inside that bridge.[7] Of the Devil's many underground dwellings, he lived in a cave at Llandulas, Clwyd,[8] while nearby at Llanrhos,[9] he occupied an old oak tree, known as The Devil's Tree, from where he jumped down onto passers-by, and drained their energy.

The tracks left by Auld Nick's coach wheels are seen on many frozen Scottish lochs in winter, while his hoofmarks are shown at an

Aberdeenshire stone circle of that name. In Yorkshire, three massive Bronze Age standing stones are known as The Devil's Arrows. Major landforms, attributed to him, include Meon Hill, Oxfordshire, which he tried to throw at Evesham Abbey.

If, on one level, Beelzebub is no more than a mischievous nuisance, in essence he is also an island-wide catastrophe, who rejects pre-historic *world affirmation*, and replaces that optimism with an all-pervasive guilt, and a view of Earth as a sin-laden misfortune.

To him, the planet is fit only for quantified exploitation, followed by a transcendental flight from its ruins. Thus we find him very much at home in that industrially ravaged area of South Staffordshire called The Black Country. From Hell Lane, Coseley, surrounded by abandoned coal mines and blast furnaces, he roamed the area, and was named at Devil's elbow, Devil's Bridge, Devil's Drumble, Devil's hole, Devil's Staircase, and Devil's Changing Room, (a quarry). He also appears, in graffiti form, at Hell Hole Yard, Walsall.[10]

A Black Country rhyme runs:

When Satan stood on Brierley Hill, And all around he gazed, He said 'I never shall again At Hell's flames be amazed.'[11]

From their triumph in the Industrial Revolution, the Satanic promoters of contemporary civilisation ignore or misinterpret the stable, cyclical repetitions of the Stone Age, (reflecting the majestic rhythm of the solar seasons and lunar phases), as embodied, though largely now unrecognised, in its surviving monuments – those abandoned theatres of good sense, that continue to offer an implied antidote to our feverish addiction to a (Devilish) linear progress.

SHEELA-NA-GIG

Given the Judeo-Christian religion's problem with the feminine gender, it is hardly surprising that portrayals of female genitalia should be perceived with horror and disgust, mingled with fascination. Such a combination of attitudes is conveyed by the numerous early medieval carved effigies, now known as Sheela-na-gig figures.

When set into the walls of churches, in Britain and Ireland, they may have been intended as grotesque warnings against the sin of lust,[1] and were shaped with a corresponding ugliness.[2] They display the dangerous nakedness of Eve's Original Sin, which led to humanity's expulsion from the Paradise garden.

One such surviving image, in Herefordshire, is known as 'the whore of Kilpeck'.[3] The figure displays her genitals as if in an invitation to fornicate.

There are more than 140 surviving statues of this type, set into the walls of parish churches, castles and monasteries.[4] Most are embedded in 12th Century Romanesque buildings, with a few now locked away in museum basements.

Yet it is the very devilishness of these sculpts that enables them to challenge their negative definition, for they were often used to banish 'the evil eye'.[5] In a face-to-face encounter with Satanic forces, the sight of a woman exposing her genitals, or a sculpted representation of such an act, was widely believed, by local 'wise women' and their customers, to effectively break Satan's spells.[6] A forthright exposure of the birthplace of every new human life was enough to trump his evil.

Such female displays probably drew upon layers of belief and practice, drawn from Antiquity, to re-emerge in Christian times, as a counter-balance to the installed orthodoxy. The Sheela who is set into the North Gate church tower, Oxford, was visited by brides on their wedding eve.[7]

As if to encapsulate an ancient concern with fertility promotion, the Sheela is often portrayed as an old female wizard, Irish *Cailleach*, with projecting ribs, a scarred face, and age-flattened breasts.[8] Her generic name probably derives from Irish *Síle ina Gío'h*, 'hag on her haunches', or hunkers,[9] rather than the misleading and arbitrary *Sheela na gig* modern title.[10] The squatting position typically described by these images, was, until recently adopted by most women in labour, and is shown by dozens of Neolithic mother goddess figurines.[11] This posture also emphasises proximity to earth, alluding to the birth of vegetation and cultivated crops, in addition to that of human babies and other animals. The sheela images draw the earth's very rock into the pregnant female image.

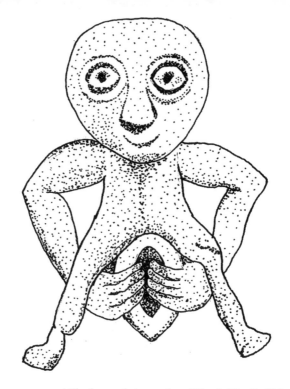

A Sheela-na-gig image from Kilpeck, Herefordshire.

The established role of many of these effigies as 'birthing stones', physically rubbed by women in labour,[12] should not exclude their topographical concern for the birth of rivers; indeed the wide-opened vulva provides a gateway to the entire fecund underworld, and all its offerings. With her hands, the woman draws aside her fleshy veil to reveal the source of life's many gifts. Here squats the entire landscape in action. She literally embodies the fabric of the planet, and displays the world as maternal 'giver'. *Sheela* is a localised mode of Greek Gaia.[13]

In Britain, Sheela derives from Celtic *síle,* alias *sulis,* appearing again as *Sulis,* the deity of the Bath springs,[14] and perhaps also the pregnant mother in harvest, as seen at Neolithic Silbury. There she is defined by Silbury's surrounding moat, which, by reflection, returns Irish *súil,* the solar 'eye' in the sky, to the goddess's water body, for another delivery or re-birth from her underworld.

Similarly, that her image is so often situated in church porches, or over doors or windows, serves literally to 'embody' entire buildings, while emphasising her own threshold functions. As with the Hindu Yoni, her vulva regulates the march of life and death through her open 'doorway'.

The scars on her face may well evoke the annual process of ploughing the fields.[15]

We should not be amazed that the Sheela is an archetype, a primordial model, that could reappear in very unpropitious circumstances. It is in the nature of archetypes to disregard prohibitions, since, in Jungian terms, they are inherited by all, from the subconscious.[16]

At Binstead, on the Isle of Wight, the sheela effigy was locally called the 'Saxon Idol'[17]. It was, and is, plainly regarded as a supernatural presence, that spans different eras, who affirms our common humanity. She conveys a sense of pleasure and gratitude, (however crudely depicted), towards each individual's mother, and also to Mother Earth, through whose stones her message is conveyed.

MARY MADONNA

What is one to make of the Blessed Virgin Mary in Goddess terms? To the pioneer feminist writer Mary Daly[1], the Blessed Virgin Mary is 'the enchained remnant of all pre-Christian goddesses, rehabilitated in defeat' and then forced into servitude within a patriarchal cult. There Mary becomes a passive onlooker, in anaemic obedience to the new authorities' demands.

Many honorific titles have been loaded upon the Madonna by Popes, both anciently, and in recent times, including that of Incorruptible Queen of Heaven,[2] Queen of the Universe,[3] Mother of God,[4] *Theotokos* (God-Bearer),[5] Mother of Disciples, and to Pope John Paul II, she is *Totuus tuus* (Latin 'totally thine').[6]

In addition, she is concerned with the redemption of every individual, yet is declared by the Church to be untainted by Eve's legacy of Original Sin. She therefore pre-dates the Biblical Genesis.

Mary and Christ as a sundial, at Inglesham church, Wiltshire.

Despite these honours and functions, in Christa Mulack's view, 'Mary serves as a flimsy projection screen, upon which archetypal goddess images are displayed as disembodied shadows.'[7] The divine is thereby tamed, and former goddesses are converted into ghosts of their former selves, yet they are greeted with touching gratitude and relief by those trapped within Catholicism's orbit.

Patricia Telesco suggests that to worship Mary together with the goddess, using your own Prayer Beads, you should 'cook rose petals in a little water in an iron pot, until they are nearly black; then add a little orris powder and rose-scented oil, and shape the beads to twice the size that you want them to be when dry. Then pierce each one with a needle, and string all together, turning them regularly until they are dry. Use them as you entreat the Goddess, energising your prayers by holding the beads.'[8] Here Mary and her rosary equipment are put to new (old) use.

But perhaps it would be better to omit the Blessed Virgin Mary altogether. She probably needs you more than you need her, now that the recently reopened great field, replete with Goddess happenings, and her varied shrines, is available for re-entry, where rediscovery of that pre-Christian divinity, in one or several of her full-blooded manifestations becomes possible again.[9]

This not to deny that there have been some surprisingly comprehensive assessments of Mary's role, as for example by the Florentine Franciscan, named Servasanctus (d. 1300). In his *Exordium*, he wrote 150 praises of Mary, to match the psalms, in which he praised Our Lady as 'the Air we breathe', and equated her to: 'light, sun, moon, cloud, air, sea, river, earth and garden'. He went on to describe her as 'bee, dove, sheep, heart' and in the cultural sphere, to 'book, mirror, vase, and column', before concluding that she was 'Mother, Queen, daughter, sister', who together provided 'the most perfect air for our pilgrimage to the Heavenly Jerusalem'.[10]

The 17th Century scholar, Hippolytus Marracci, in his *Polyanthea Mariana*, traced four more medieval instances of Mary being referred to as air, but with the 19th Century poet, Gerard Manley Hopkins, we return to the familiar reduced version of an airy Mary, a personal comfort-giver, a child's blanket to protect against the sin-infested world, as the following excerpts from his long poem of 1883 may indicate:

> I say that we are wound
> With mercy, round and round
> As if with air. The same
> Is Mary, more by name.
> Be thou then, O thou dear
> Mother, my atmosphere
> My happier world, wherein
> To wend, and meet no sin.
> World Mothering air, air wild,
> Fold home, fast hold *your* child.[11]

So Mary returns to ameliorate the burden of sins that were not of her making. Her brief excursion into Servasanctus' territory proved short-lived. One can but feel sorry for her followers in a religion that has generally found such slight use for the full feminine range of attributes, both sacred and practical. May Mary meanwhile do her best to ease the unnecessary hardship of her many followers.

INDIVIDUAL AND DEITY

Kathy Jones is a leading figure within the modern goddess movement in Britain and beyond. Michael Dames asked her to recount her spiritual and creative passage over the last four decades.

My journey with Goddess began in the late 1970s when I first moved to Glastonbury as part of my spiritual passage. After being here for a few months a group of us set up a women's consciousness-raising circle. We met every fortnight for an evening of conversation based around what was known then as the Seven Demands of the Women's Liberation Movement. Each woman talked in turn about her experiences as a woman in the world and we listened to each other. We shared our understandings and insights.

We met for about two years and as part of this process we began to take part in women's political actions. One of the group, Stephanie Leland, created a magazine called *Women for Life on Earth*, which was inspiring. We created a small magazine called the *Glastonbury Thorn*, which included some of my first forays into writing about Goddess, as I was beginning to learn about Her.

I had always been a spiritual seeker since childhood but until my Saturn Return my seeking had always been based on He, Him and His, as I knew nothing other. I had never been taught about She, Her and Hers. Through the women's group I began to read ancient Goddess myths from around the world and learned to understand them a little. Before this time they had been meaningless stories.

It was the time of Greenham Common and the Women's Protests and Peace Camp, which were based there. Many women went from Glastonbury, just a small country town, to join in the protests against American Nuclear Cruise Missiles being brought into Britain, to Greenham Common. We wanted to protect our land and our children's future from these deadly weapons of mass destruction. I visited Greenham several times and took part in the *Embrace the Base* action, in which a circle of women 20 miles long, surrounded the army base and decorated the green chain link fence with photos of our children and other decorative pieces. We sang our songs of life; we danced and did a few crazy things. It was very empowering.

I remember going to Greenham one time and standing at twilight in the woods outside the perimeter fence. We all held candles in the darkness and sang Goddess songs to the land and to the soldiers and policemen who were on the inside of the perimeter fence, protecting the silos which would soon house the missiles. We sang,

> You can't kill the Spirit
> She is like a mountain
> Old and strong,
> She goes on and on and on…
> (by Naomi Littlebear Morena)

This was the first time that I can remember singing about the Spirit as She. Before then it was OK for the Earth to be She – our Mother Earth, but Spirit had always been He. This was a moment of liberation for me. I understood then that Spirit is She, not in a relational way to He, but as Herself, as She.

Two women from Glastonbury, Liz Beech and Jill Booth, left their families and went to live at Greenham in the mud. I wanted to celebrate them and what they were doing. I had recently read the Greek myth of Demeter and Persephone, which describes the abduction and rape of Persephone and her descent into the underworld of Hades/Pluto. I saw this as an allegory for what was happening at Greenham. The industrial military complex that creates weapons of war needs to initiate and perpetuate wars to fulfil their need for profits and they need to be continuously fed. I saw this complex as lying at the core of the underworld of Hades. I wrote a short play and asked people to come and participate in the play at the Glastonbury Assembly Rooms, a local community centre. We gave a one-night performance which was very successful and inspiring.

In the following year I read *Inanna, Queen of Heaven* by Diana Wolkstein and Samuel Noah Kramer. This is a poetical translation of the earliest love story in the world, from Sumer, which is modern day Iraq. It was found written on clay tablets and dates from about 2000 B.C.E. The language is beautiful and I really fell in love with the story. Like the myth of Demeter and Persephone it tells of the Descent of the Goddess Inanna into the underworld, but this time She goes to visit a much earlier Goddess, Erishkigal/Pluto, her grandmother. I decided to create another performance based on this myth and this time with music and songs and a company of about 50 people. The performance again was a great success and we were all inspired by the experience of seeing and hearing the words and songs of the ancient Goddesses on stage.

As my experiences with creating dramatic productions grew I started a 13-year journey rewriting patriarchal myths and bringing the Goddesses hidden within them into focus. I also began to understand that when these divine beings appeared on stage something numinous also occurred and participants and audiences began to experience something divine. We had begun our journey with sacred drama, rather than just drama per se.

One of the early plays *The Beauty in the Beast* I wrote with my friend Leona Graham. It was inspired by Ariadne the Snake Goddess of ancient Krete and the legend of the Minotaur, and so Ariadne Productions, our sacred drama company, was born. Over the next years I wrote sometimes with others, sometimes alone, and directed and performed in one or two intense sacred dramas each year. The story of this journey is told in my book *On Finding Treasure: Mystery Plays of the Goddess* (Ariadne Publications). It is an exploration of the powers of sacred drama, the embodiment of Goddess, and the reworking of ancient Goddess myths to bring them into the modern age.

It was after creating a few of these dramas that I realised that some of the Goddesses of the world had seriously begun to speak to me through these ancient myths and in a short while more directly and personally. I remember sitting writing at my desk in my bedroom. On the side wall was a lovely poster of Green Tara, the Tibetan Goddess of Compassion. She had been given to me 10 years earlier by an ex-lover. She was there because the image had lovely colours and matched the wallpaper. I knew nothing about Her. As I sat writing I heard a female voice coming over my right shoulder from the poster, and this was not something that happened to me.

She said, *Look at me.*

I turned to look at the wall and the poster. What I saw astonished me. It was as if Green Tara had become 3-D and was sitting on the wall, the same size as Her image in the poster.

She said to me, *Find out who I am and write a play about me.*

I was and am someone who hears the instructions of intuition and divinity, and follows these inspirations, sometimes with resistance. But essentially, I hear and obey. So I found some books about Green Tara and began to meditate upon Her using Her mantra, *Om Tara Tuttare Ture Svaha.*

I found that I already knew Her. Somewhere deep inside, in my bones, I knew who She was. I knew Her qualities and energies. I had known Her deeply before, in another life, where I had loved and served Her. I explored Her story and then wrote a play about Her with another writer, Peter Davis. The drama was performed for three nights and it was through these performances that I met my life partner, Mike Jones, who came to play music for us with other musicians.

Goddess brought Mike and I together along different life paths to meet in a place where we had known each other before, once as monks in Tibet. Our ancient connection was sealed over the synchronous sounding of Tibetan bells. When we came together we also found out that we had followed similar mythic paths that included deep experiences of Green Tara, of Ariadne of Krete, and of Pan. My path came through sacred drama and Mike's journey was through playing music in a psychedelic rock band. Goddess brought us together in the right moment of remembering, for which we are both deeply grateful.

This process of exploration and the creation of sacred dramas continued with different Goddesses from around the world. It was as if each one called to me and asked me to remember Her true story, which came from pre-patriarchal times. It was a wonderful Goddess journey. By the late 1980s I realised that I had written about the Goddesses of many different lands but not about Goddesses in Britain, because of course at that time Britain was a Christian country and we didn't have Goddesses. As soon as I thought this I realised it had to be wrong. Goddess is everywhere. She had to be here too. We just had to find Her again and the way to find Her here in my own land was to look at the earth and the landscapes around us. Because Goddess is our Mother Earth and She reveals Herself to us through the shapes, forms and names of the land, through the seasons of Her nature and through the weather. She is also hidden within folklore and legends.

Mike and I looked at ordnance survey maps of the Glastonbury area and soon found Goddess revealed here as a great landscape Swan who is the Maiden, as a Lover, as a Great Mother and as an aged Crone. She showed herself to us in the land.

I looked for other Goddesses in Britain and after thinking there weren't any, within a short time I found literally hundreds of Goddesses hidden within the names given to landscapes, rocks and mountains, such as the Hag of Bheara, the Paps of Jura. I read and was deeply inspired by Michael Dames' books, *The Silbury Treasure* and *The Avebury Cycle*. I found Goddess names in legends and stories.

In short succession I wrote and self-published first *The Goddess in Glastonbury* and in 1990 the first edition of *The Ancient British Goddess: Goddess Myths, Legends, Landscapes and Present Day Revelations* (Ariadne Publications), which I updated 10 years later.

Through the late 1980s and 1990s I taught many day and weekend workshops on Goddess and created many ceremonies celebrating the seasons of Her nature. I helped to co-found the Library of Avalon, the Bridget Healing Wing and the University of Avalon, later called the Isle of Avalon Foundation. These spiritual, educational and healing ventures were mainly located in the Courtyard of the Glastonbury Experience, and still exist today. This all happened due to the generosity of its Dutch owner Helene Koppejan and Barry Taylor.

I led hundreds of people through the great Goddess Labyrinth that encircles the slopes of Glastonbury Tor. I worked with many spiritual groups who came on pilgrimage to Glastonbury from Europe and the USA, taking them on Goddess Pilgrimages through the Glastonbury landscape and to Avebury and Stonehenge. Marion Zimmer Bradley had published her novel *The Mists of Avalon,* which spoke to thousands of women worldwide about the Goddess and the mysterious and ancient world of Priestesses, who might be contacted here in Glastonbury. Over the years I worked with many women who came to Glastonbury in search of the ancient Priestesses of Avalon.

In 1995 I was inspired to create the first Glastonbury Goddess Conference, with the help of a friend, Tina Redpath. We wanted to create a Goddess event which celebrated British Goddesses and which encouraged British Goddess women and men to be creative as ceremonialists, writers, artists, craftspeople, poets, musicians and performers. At Lammas 1996 we held the first Glastonbury Goddess Conference, with many Goddess women and a few men coming from Britain and all over the world to celebrate Goddess. It was a great and challenging success, not everyone wanted to support us, women and men, but we continued anyway. The Goddess Conference has continued annually from that time and in 2015 is in its twentieth year. See www.goddessconference.com

The Goddess Conference has developed over the years to become a six- and nine-day Pilgrimage into the Mysteries of Avalon, home of Goddess. As it grew the need to have priestesses who could hold the energy of the Conference and all the people attending it, soon became evident. At first I had invited women who had taken part in the sacred dramas to help priestess the Conference, but soon the need to have truly dedicated priestesses became obvious.

I decided to set up the Priestess and Priest of Avalon Training, for those who wished in this life-time, to dedicate themselves once again to the Lady of Avalon, Goddess of the Sacred Isle of Avalon. The Lady of Avalon is an ancient Goddess whose name and presence speaks to many people, yet she is not named in any records. She is a new-found Goddess, reclaimed in the Priestess of Avalon Training as the focus of our love and dedication. I was very aware of the glamour attached to this title and the hubris that I might be committing by taking this title on again and daring to teach such a training. After thinking about it for two years I set up the first Priestess/Priest of Avalon Training in 1998 for all those women and men who wished to learn how to serve Goddess in this way. It seemed important that men too should have this opportunity to show their love and devotion to Goddess, as well as women.

The first Millennial Priestesses of Avalon dedicated themselves to the Lady at Mabon in 1999. It seemed so right that this happened at the turning of the new 21st Century. After a while it became clear that the process of learning was much deeper than I knew at first and in a short while the training expanded to become a three-year, Three Spiral Training. The First and Second Spirals are now taught by my sister Priestess of Avalon, Erin McCauliff, and I teach the Third Spiral as well as further Spirals, which include the Advanced Priestess Training, Priestess Healer and Priestess Enchantress.

I wrote and published other books – *In the Nature of Avalon: Goddess Pilgrimages in Glastonbury's Sacred Landscape* and *Priestess of Avalon, Priestess of the Goddess: A Journey of Transformation within the Sacred Landscape of Glastonbury Avalon.* This latter book is now the foundation book for the Priestess Training. I wrote a book about my experience with breast cancer in 1995/6, *Breast Cancer: Hanging on by a Red Thread* and also *Chiron in Labrys: An Introduction to Esoteric Soul Healing*, which I also teach.

Over the years Mike and I travelled to many ancient Goddess sacred sites around the world, from Krete, to Greece, to Anatolia, Sweden, Sapin and all over Brigit's Isles. In the early days everywhere we went the Goddess Temples we visited were in ruins. One year when we were on holiday we went for a day visit to Mount Olympus. As the bus approached the mountain a dense mist arose, just like approaching Avalon. It cleared suddenly and we came through to see the many beautiful peaks. At the foot of the mountain are the ruins of Dion, an ancient city. In the city are the ruins of Temples to Isis, Demeter and Hera. It was here that I felt overwhelmed by grief.

We had visited so many ancient ruined Goddess Temples in the world and here we were at another one. I couldn't bear it. It was then that I resolved that when we got home I would create a living, modern-day Goddess Temple in my hometown, in Glastonbury. It would be a Goddess Temple which was not forgotten and in ruins. I would create a living present day Goddess Temple. On our return home I called a meeting of people I thought might be interested in helping to create a Goddess Temple.

We began by hiring a room in the Courtyard of the Glastonbury Experience in the High St, for a few days at the time of Beltane, one of the seasonal festivals. We decorated it as a Goddess Temple with colourful materials. We made a shrine and placed one of the Conference wicker Goddesses, made by sculptor Foosiya Miller, inside it. We held a beautiful celebration for Beltane. We offered everyone strawberries and champagne and had a wonderful time. We held the space open for three days for people to come and pray, meditate, sing, dance and love the Goddess. We received donations from people who came, which helped pay for the room. When the three days were over we took everything down. We had such a good time that we decided to do it again at the following seasonal festival, and hired another room and created another Temple.

We continued to do this for the next eighteen months creating Temple spaces for each of the eight seasonal festivals of the year.

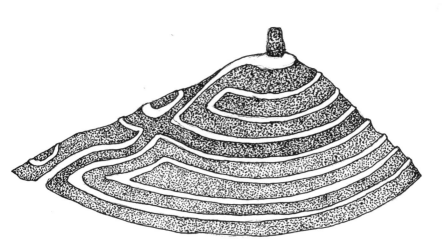

Glastonbury Tor, Somerset, with its church tower.

In 2001 a more permanent space became available within the Courtyard and we took the plunge and rented it, to set up as a permanent Goddess Temple. With great excitement we opened to the public at Imbolc 2002. We registered ourselves as a Place of Worship, not realising until a while later that this was in fact, a herstoric event. We are the first and so far only (in 2014) formally registered indigenous British Goddess Temple in Brigit's Isles and the only officially recognised Goddess Place of Worship in Europe for perhaps 1,500 years.

Since we opened thousands of people have come through the door to visit the Goddess Temple, from all over Brigit's Isle and from many of the countries of the world, from Australia, Japan, the USA, Russia and Europe. Our small Goddess Temple is open every day of the year from 12pm to 4pm. The Temple is cared for by Temple Melissas – bee priestesses, who give their time to serve the Queen Bee, our Lady of Avalon, who is our central Goddess. We are financed mainly by donations of time, energy and money from our Temple Melissas and Temple Madrons, who give the Temple regular monthly donations, as well as donations from visitors, and also from the Priestess Training courses we teach.

After a few years we had so many people coming to our public seasonal ceremonies that we were in serious need of a bigger Temple space to accommodate all those wishing to participate. In 2008 we bought our larger Goddess Hall, which is located in Benedict Street, which we use for our seasonal ceremonies and for teaching. We decorated the Hall as another Goddess Temple with glorious paintings and decorations, including the wonderful new mural of the ancient Isle of Avalon, on the front wall by artist Jonathon Minshull. Our Temple teachings now include Priestess of Avalon, Priestess of Rhiannon with Katinka Soetens and Priestess of Bridget with Marion Van Eupen, Trainings, as well as other Goddess workshops.

In 2013 we set up our Goddess Temple Gifts shop in the Courtyard, which gives priestesses and Goddess artists and craftspeople the opportunity to earn their living from their Goddess-inspired work. It's a beautiful shop managed by artist Sue Quatermass. Profits from the shop go to support the Temple.

We have grown organically from small beginnings, step by step, inspired by the Lady of Avalon. We are now looking for larger Goddess Temple premises and funds to pay for them. We would like to have several larger and smaller Temple spaces, teaching spaces, a Goddess café, a bakery, a dedicated Goddess Healing Centre and many other ventures, where we can love and serve the Lady in everything that we do.

At the 2004 Goddess Conference we created the Flame of Avalon, the Flame of the Lady which shines out into the world, bringing Her Violet Flame once again into all the darkened places of the world. The Flame of Avalon was created from the Flame of the 2004 Goddess Conference, which was lit from the sun and kept alight throughout that Conference; the Flame of Hiroshima, lit from the embers of the fire that raged and kept alight for 50 years; Brigid's Flame from Kildare in Ireland brought to us by the Brigidine Nuns, Sisters Mary and Sister Rita; Brighde's Flame from the Hebrides; the Madonna Ministry Flame; and the Children's Flame from the USA. The Flame of Avalon has been kept alight in the hearts and minds of many people, including a circle of Priestesses of Avalon, who hold alight the Flame for 24 hours at a time in honour of the Lady and Her radiating light. Since that time several other Goddess Flames have been added to the original Flame, including Flames from South Africa, and Gobekli Tepe in Turkey, one of the earliest Goddess sites found so far.

Visitors to the Glastonbury Goddess Temple are invited to light a candle from the Flame of Avalon and to take it with them to light their own candles at home and with their own Goddess groups, holding the Flame of Avalon, the Light of the Goddess in the wick of the candle. Thus the Lady's light and love shine out into the world more and more every day, bringing Her back into our consciousness, into our human awareness once again.

Kathy Jones, 2015

SOME SACRED GEOGRAPHIES

Within the confines of two offshore European Islands we may be unaware that a Sacred Geography, of one sort or another, is recognised and experienced, on every continent. These continental lands are typically regarded as divine living entities. They provide the foundation upon which many cultures trace their origins and store their accumulating collective memories, while also involving these deities of the land in celebrating seasonal rites, and the threshold events of each individual's life cycle.

E.J. Eitel, a 19th Century Christian missionary to China, discovered to his astonishment that 'The Chinese look upon Nature not as a dead inanimate fabric, but as a living, breathing organism. They see a golden chain of spirited life running through every form of existence, and binding together, as in one body, everything in Heaven and Earth. They call this life-force Chi; it manifests in the atmosphere, in the earth, and within the bodies of animals, people and plants.'

Moreover, all of China, south of the Great Wall, was organised into a unifying sacred geometry, involving five sacred mountain peaks. These peaks were seen as charged with supernatural power. Invested with layers of mythic meaning, they served as focal points in orientating the surrounding country, so enabling the population to appreciate its

Earth, moon, sun, ocean and stars, the union on which our lives depend.

overall structure, and, beyond that, to help explain his or her place in the cosmic schema.

The sense of a sacred network was also greatly enhanced by processional ways and pilgrimage routes, repeatedly trodden over many generations. Thus, rather than being regarded as abstract *concepts*, sacred geographies were participated in, and *performed* as a vital aspect of communal life. This is equally true among both sophisticated and supposedly primitive societies.

Hunter-gatherer groups of Australian aborigines, established there for c. 30,000 years, follow their 'Dreamtime' original ancestors, along with the first animals, who were believed to have emerged from within the earth, and to have created its rock-stack and waterhole features during their first perambulations.

These first routes, known as 'song-lines', were then re-walked by each subsequent generation, in a concert of recreation, involving rites of passage for young adults.

Likewise across medieval Europe, waves of crusaders made the hazardous journey to Jerusalem. There they hoped to find a divine understanding of their earthly existence, by relating to the climax of Christ's life story.

One could go on to outline the rich variety of sacred geographies that have given structure to human life in the Americas. Suffice to say that my attempt, in this book, to recover evidence for a Sacred Geography of Britain and Ireland is not an eccentric enterprise, but rather an invitation to our island's inhabitants, present and future, to recognise the remnants of our own sacred geography, and in doing so, to re-join the mainstream of human culture on our beloved planet Earth.

If we believe our own words, we should not ignore the often spoken figure called GE. She still provides the first syllable of modern GE-*ography,* GE-*omor-phology, and* GE-*ology. From this trinity, when married to her offspring, sky and ocean, she created a lasting, fertile synthesis. Today, we might try to sustain her well-earthed sacred legacy.*
Best Wishes, Michael

Acknowledgements
First and foremost I gladly recall substantial help from my wife Judith Anne Dames. Secondly I wish to thank Mark Pilkington and Jamie Sutcliffe at Strange Attractor Press, Hilary Mortimer for her editorial work, Natalie Kay-Thatcher for her illustrations and Matthew Shaw for introducing my book to Strange Attractor. I am grateful to the staff of Birmingham University Library, and the City of Birmingham Library. I also need to say thank you to Joan Marler, in California, Kathy Jones of Glastonbury, Mary Condren in Dublin, and to Thomas Neurath for launching me on my way, so long ago.

NOTES, SOURCES AND REFERENCES

INTRODUCTION

1. Publius P. Statius, *Thebaid*, 1st Century A.D. (transl.) Mozley, J.H. (New York: Longman, 1928) p. 295 ff.

2. Hesiod, *Theogony*, c. 700 B.C. (transl.) Evelyn-White, ,(Cambridge, Massachusetts: Harvard University Press, 1914) pp. 116-138.

3. Smith, William, A *Dictionary of Greek and Roman Mythology*, Vol. iii (London: John Murray, 1845).

4. Ovid, *Fasti*, IV, 629. And Adkins, Lesley and Adkins, Roy A., *A Dictionary of Roman Religion* (New York: Facts on File, 1996).

5. Harrison, Jane Ellen, *Prolegomena to the Study of Greek Religion* (Princeton, New Jersey: Princeton University Press, 1903) pp. 77-85. See also Krates, 2nd Century A.D.

6. Dames, Michael, *The Silbury Treasure, The Great Goddess Rediscovered* (London: Thames and Hudson, 1976) pp. 177-9.

7. *Hymn to Gaia*, composed 7th-4th Centuries B.C. (transl.) Evelyn-White, Hugh G., Loeb Classical Library, Vol. 57 (London: Heinemann, 1914). Hymn No. 30, from ll.1-16. And by Roth, from *Hymn to Gaia*, 7th-4th Century B.C. (From Hymn no 11).

I. LIFE WORLD

• Ceridwen's Legacy

1. Gruffydd, ,*Bulletin of the Board of Celtic Studies*, Vol. 7 (1993) pp. 1-4.

2. Rhys, John, *The Hibbert Lectures* 1886 (London: Williams and Northgate, 1888) pp. 89-92.

3. Ford, E.K., *A Fragment of Hanes Taliesin by Llewellyn Sion* in Etudes Celtique, Vol. 14 (1975) pp. 451-460.

4. Guest, Lady C., *The Mabinogion*, 2nd edit. (London, 1877) pp. 471-494.

5. Dames, Michael, *Taliesin's Travels, a Demi-god at Large* (Avebury: Heart of Albion Press, 2006).

• Salmon-Woman

1. Hartland, Edwin Sidney, *Folklore*, Vol. 16 (1905) pp. 535-7.

2. Dames, Michael, *Mythic Ireland* (London: Thames and Hudson 1992) pp. 168-9 and pp. 190-1.

3. Gray, (ed.) *Cath Maige Tuired: The Second Battle of Mag Tuired* (Dublin: Irish Texts Society, 1982) p. 23.

4. *Lebor Gabála Érenn*, vol. III (Dublin: Irish Texts Society, 1940) p. 23.

5. Cihar, Jiri, *Freshwater Fish* (London: Octopus Books, 1977) pp. 60-62.

• Lady of the Plants

1. Freyja as 'Lady': Branston, Brian, *Gods of the North* (London: Thames and Hudson 1980) p. 133.

2. Freyja as *syr*, 'sow', ibid., p. 132.

3. Britain's local plant names gathered from Grigson, Geoffrey, *The Englishman's Flora* (London: Phoenix House, 1987) pp. 439-469.

• **Snake Deities**

1. A term invented by King Henry VIII.

2. Book of Genesis, Ch. 3, Verse 20.

3. Ibid., Ch. 3, verses 2-7.

4. Ibid., Ch. 3, verses 21-24.

5. Ekwall, E., The Concise Oxford Dictionary of English Place-Names (Oxford: OUP/Clarendon Press, 1960).

6. The Lambton Worm in Jacobs, J., *More English Fairy Tales* (London: D Nutt, 1922) pp. 196-203.

7. The Roxborough Worm in Parkinson, D., *Mysterious Britain and Ireland* (www.mysteriousbritain.co.uk/) 2016.

8. The Meister Stour Worm, often spelled Mester Stoor. See Westwood, Jennifer and Kingshill, Sophia, *The Lore of Scotland* (London: Random House, 2012) pp. 273-5.

9. Simpson, J. *Folklore of Sussex* (London: Kindle edition, 2009) pp. 16-17, and pp. 35-9.

10. The Backworth hoard, Walls End, North Tyneside; it is now in the British Museum.

11. The Snettisham Hoard, 100 B.C. See *Current Archaeology*, May, 2006, Vols. 125-6.

12. Dames, Michael, *The Avebury Cycle* (London: Thames and Hudson, 1977) pp. 141-2.

13. Ibid., pp. 85-6.

14. Reid, John, The Rotherwas Serpent, Herefordshire, *Current Archaeology*, (11th Oct., 2007).

15. Hamilton, Ross, *The Mystery of the Serpent Mound: In Search of the Alphabet of the Gods* (Mumbai: Frog Books, 2000).

16. MacKillop, James, *Dictionary of Celtic Mythology* (Oxford: OUP, 1998).

17. Gordon-Cumming, Constance F., *Nature*, Vol. 20, Issue 506 (1879) p. 242. Gordon-Cumming, Constance F., *In the Hebrides* (London: Chatto and Windus, 1883).

18. Trevelyan, Marie, *Folk-lore and Folk-Stories of Wales*, 1863 (London: Eliot Stock, 1909).

19. Milford, H. (ed.) *Oxford Book of English Verse of the Romantic Period* Oxford: The Clarendon Press, 1951), pp. 537-540.

20. Keats, J., in Poetical Works of John Keats, Garrod, H. W. (ed.) (London: Oxford University Press, 1959), pp. 161-178.

21. Wordsworth, W. *from Lines Composed a Few Miles Above Tintern Abbey*, 1798, in *The Poetical Works of William Wordsworth* (Oxford: OUP, 1976) p. 173.

22. Cook, E. T. and Wedderburn, A. (eds) *The Works of John Ruskin*, Vol. 4 (London: Cambridge University Press, 1903-12) p. 182

23. Cup and Ring marks, Bradley, R., *The Archaeology of Rock Art* (Cambridge: Cambridge University Press, 1997) pp. 130-145.

• **Holy Cows**

1. Hemming, J. in *Folklore*, Vol. 113 (1971-2).

2. Vischier, et al. A viable herd of genetically uniform cattle. In *Nature*, Vol. 409, Jan. 2001, p. 303.

3. McKenna, T., *Food of the Gods,* pp. 100-116. And *Rig Veda,* 4. 28.1.6.

4. Norayen, Jha Dwendra, *The Myth of the Holy Cow* (London & New York: Verso Books, 2002).

5. *New Larousse Encyclopedia of Mythology*, 1970, p. 87.

6. Herodotus, *The Histories* Book 1.2. and Ovid, *Metamorphoses*, p. 264.

7. Wentworth, H., *Dictionary of American Slang* (New York: Thomas Y Crowell, 1962) p. 147

8. Joyce, P.W., *The Wonders of Ireland* (London: Longman's Green, 1911).

9. Dames, Michael, *Taliesin's Travels: A Demi-god at Large* (Avebury: Heart of Albion Press, 2006) pp. 188-9.

10. Wilde, J.C.F., *Ancient Legends of Ireland* (London: Chatto & Windus, 1888) pp. 102-3.

11. *New Larousse Encyclopedia* of Mythology, op. cit., pp. 55-6.

12. Palmer, Roy, The Folklore of Shropshire (Hereford: Logaston, 2004) p. 211.

13. *New Larousse Encyclopedia of Mythology*, op. cit., p. 248.

• **Bees: Poetry, Magic and Regeneration**

1. Roud, Steve, *The Penguin Guide to the Superstitions of Britain and Ireland* (London: Penguin Books, 2006).

2. Ransome, H., *The Sacred Bee in Ancient Times* (Dover Publications, New York, 2004) p. 229.

3. Gimbutas, Marija, *The Gods and Goddesses of Old Europe* (London: Thames and Hudson, 1974), p. 182.

• **The White Lady in Britain**

1. Trevelyan, Marie, *Folklore and Folk-stories of Wales* (London, 1909).

2. Beck, J.C., The White Lady of Great Britain and Ireland, *Folklore,* Vol. 81, (1970) pp. 292-302.

3. Udal, John Symonds, *Dorsetshire Folk-lore* (1922) p. 74.

4. *New Larousse Encyclopedia of Mythology* (London and New York 1968) p. 170.

5. Hayward, L.M., Shropshire Folklore, in *Folklore*, Vol. 49, p. 239.

6. Dames Michael, *Mythic Ireland* (Thames and Hudson, London, 1982), p. 181.

7. Hartland, E.S., *English Fairy and Folk Tales* (W. Scott, London, 1890) pp. 219-22.

8. Dames, Michael, *The Silbury Treasure* (London, Thames and Hudson, 1976).

9. Trevelyan, op. cit. 1909, pp. 200-1.

10. Ibid., p. 211.

11. Gwynn-Jones, T., Welsh *Folk Lore and Folk Custom* (Methuen & Co., London, 1930) p. 38.

12. Tongue, Ruth L., *Somerset Folklore* (London: Folklore Society, 1965).

13. Ibid., p. 120.

14. Ceredig Davies, Jonathan, *Folk-lore of West and mid-Wales* (Aberystwyth: Welsh Gazette, 1911).

15. Briggs, K. M., *The Folklore of the Cotswolds* (London: B.T. Batsford, 1974) p. 143.

16. Palmer, R., *The Folklore of Warwickshire* (London: B.T. Batsford, 1976) p. 78.

17. Bartrum, Peter C., *A Welsh Classical Dictionary: People in History and Legend Up Until About A.D. 1000* (Aberystwyth: National Library of Wales, 1993) p. 318.

18. Dames, Michael, *Taliesin's Travels* (Loughborough: Heart of Albion Press, 2006), pp. 34-5.

19. Ibid., p. 35

20. Ibid., p. 35.

21. Ibid., p.186.

22. Fittock, Matthew J., *The Pipe-Clay Figurines from Roman London*, University of Reading, The Roman Archaeology Conference, March 27-30, 2014.

23. *Taliesin's Travels*, pp. 188-9.

24. *Montgomery Collections*, Vol. 35 (1908-9) p. 183.

25. Wikipedia, Philippine Urban Legend, July 29th 2012.

26. Doolittle, Hilda, *Tribute to the Angels* (London: OUP, 1944).

• **The Divine Sun**
1. *New Larousse Encyclopedia of Mythology*, 1968 edition. p. 413.
2. Ibid., pp. 333-5.
3. Trikunas J., Of Gods and Holidays: the Baltic Heritage (Lithuania: Tverme, 1999) pp. 75-7.
4. New Larousse Encyclopedia of Mythology, op. cit. p.308
5. Monaghan, P., Encyclopedia of Goddesses and Heroines, 2 Vols, (Santa Barbara: Greenwood Publishing, 2014) p. 212.
6. Ibid., p. 369.
7. Tag Archive, Basque Mythology, 2014.
8. Branston, B., Gods of the North (London: Thames and Hudson, 1980) pp. 65-72.
9. Ibid., p. 67.
10. Dictionary of the Irish Language (Dublin: Royal Irish Academy, 1980).
11. O'Rahilly, T. F., Early Irish History and Mythology (Dublin: Dublin Institute for Advanced Studies, 1946) p. 293.
12. Dames, Michael, Mythic Ireland (London: Thames and Hudson, 1992) pp. 62-3.
13. Ibid., p. 246.
14. Ibid., p. 251.
15. Stokes W., *Lives of the Saints from the Book of Lismore* (Oxford: Clarendon Press, 1890) and Bowen, E. G., The Cult of St Brigit, in *Studia Celtica*, Vol. 8 (1973).
16. Stokes, W., op. cit., p. 184.

II. THE DANCING GODDESS

• **An Archaeo-Mythic Grammar**
1. Flew, A., *Dictionary of Philosophy* (London: Macmillan, 1979) p. 278, p. 306. See also Aristophanes, *The Clouds*, a satirical comedy featuring Socrates, written and staged in 423 B.C.
2. Ibid., p. 83-85.
3. Thomas, Julian, *Archaeology and Modernity* (London: Routledge, 2004) p. 234.
4. Russell, Ian Alden (ed.) *Images of the Past: Images, Representations and Heritage* (New York, Springer-Kluwer, 2006) pp. 18-19.
5. Renfrew, Colin, *The Ancient Mind* (Buffalo: State University of New York, 1994) pp. 3-4.
6. Thomas, J., op. cit., p. 248.
7. *Oxford English Dictionary*.
8. Mitchell, W. J. T., *Iconology: Image, Text, Ideology*, (Chicago: University of Chicago Press, 1987), p.p.1-5.
9. Lippard, Lucy R., *Looking Around*, in Mapping the Terrain: New Genre Public Art, Lacy, Suzanne (ed.) (Seattle: Bay Press, 1994) pp. 128-9. See also her *Overlay: Contemporary Art and the Art of Prehistory* (New York, Pantheon Books, 1983).
10. Lipner, Julius, *Hindus: their Religious Beliefs and Practices* (Manchester: Manchester University Press, 1994).
See also: Fisher, Nora (ed.) *Mud, Mirror and Thread: Folk Traditions of Rural India* (Boston: Kluwer Academic Publishers, 1992).

• **Marija Gimbutas' Gift**

Gimbutas, Marija, *The Gods and Goddesses of Old Europe: 7000 to 3500 BC: Myths and Cult Images* (Oakland, California: University of California Press, 1974).

Gimbutas, Marija, *The Language of the Goddess: Unearthing the Hidden Symbols of Western Civilisation* (New York: Harper and Row, 1989).

Gimbutas, Marija, *The Civilisation of the Goddess: The World of Old Europe* (New York: Harper, 1991).

• **The Silbury Monument as a Work of Art**

1. *Silbury Hill*, 1970-1, sculpture by Richard Long. 'A line the length of a straight walk from the bottom to the top of Silbury Hill'.

2. *Proceedings of the European Society for Aesthetics, Volume 2* (2010) pp. 491-515.

3. Smithson's Spiral Jetty sculpture was constructed in Utah, USA in 1970.

4. Kant, Immanuel, *Critique of Judgement*, 1790, paras 23-28.

5. Gormley, Antony, in Searle, A., *A Field for the British Isles*, in Hayward Gallery pamphlet, London, 1996.

6. Jim Leary speaking in Symonds, Matthew, The Many Faces of Silbury Hill, *Current Archaeology*, Issue 293 (August 2014) pp. 13-18.

7. See Silbury at Play: the Goddess in Folk Drama, in this book.

8. Symonds, Matthew, op. cit., p. 15.

9. Dames, Michael, *The Avebury Cycle* (London: Thames and Hudson, 1977) pp. 110-171.

10. Piggott, Stuart, *West Kennet Long Barrow Excavations, 1955-6*, Ministry of Works Archaeological Reports No. 4 (London: HMSO, 1962) p. 61.

11. Mookerjee, Ajit, *The Ritual Art of India* (London: Thames and Hudson, 1991).

12. Symonds, Matthew op. cit., p. 17.

13. Long, William, *Wiltshire Archaeological and Natural History Society*, Vol.4 (1858).

14. Dames, Michael, *The Silbury Treasure* (London: Thames and Hudson, 1976).

15. MacNeill, Máire, *The Festival of Lughnasa: A Study of the Survival of the Celtic Festival of the Beginning of Harvest* (London: Oxford University Press, 1962).

16. Jackson, John Edward (ed.) *Wiltshire: the Topographical Collections of John* Aubrey, 1659-70 (Devizes: Wiltshire Archaeological and Natural History Society, 1872).

17. Jencks, Charles, *What is Post Modernism?* (London: Academy Editions, 1996).

18. Klee, Paul, in Mannheim, Ralph. (transl.) *The Thinking Eye, 1914* (Basle: Schwabe, 1956, English edition 1961).

19. Dewey, John, *Art as Experience* (New York: Putnam, 1934).

20. Came, Daniel (ed.) *Nietzsche on Art and Life* (Oxford: OUP, 2014).

21. Tolstoy, *Leo*, (transl.) Aylmer, Maude, *What is Art?*, 1896 (Manchester: Albert Broadbent, 1900).

22. Jung, Carl, Portrait of the Visionary Artist, *Time* magazine, February 14th, 1955.

23. Coleridge, Samuel Taylor, in *Biographia Literaria*.

• **Silbury at Play: The Goddess in Folk Drama**

1. Heidegger, Martin, *Poetry, Language, Thought* (transl.) Hofstadter, Albert (New York: Harper Perennial, 1975), containing The Origin of The Work of Art, pp. 17-87.

2. Field, David and Leary, Jim, *The Story of Silbury Hill* (London: English Heritage, 2010).

3. *Shorter Oxford English Dictionary*; mythical 'having no foundation in fact'.

4. Kisiel, Theodore, et al, *Heidegger's Way of Thought: Critical and Interpretative Signposts*

(London: Bloomsbury Academic, 2002) p. 198.

5. Ibid., p. 198.

6. Heidegger M., op. cit., pp. 179-180.

7. Dames, Michael, *The Silbury Treasure* (London: Thames & Hudson, 1976) p. 44.

8. Field and Leary, op. cit.

9. Eliade Mircea, *Myth and Reality* (New York: Harper Torch Books, 1959) pp. 198-9.

10. Heidegger, op. cit. pp. 179-180.

11. Ibid. p. 170.

12. Ibid. p.181.

13. Ibid. p. 43.

14. Hein, Norvin, *The Encyclopedia of Religion*, Vol. 5 (Chicago: University of Chicago, 1995) pp. 550-554.

15. Stutley, James and Margaret, *A Dictionary of Hinduism: Its Mythology, Folklore and Development, 1500 B.C.-1500 A.D.* (London: Routledge & Kegan Paul, 1st Edition 1977) p. 161.

16. Dames, Michael, *The Avebury Cycle* (London: Thames & Hudson, 1st Edition, 1977).

17. Heidegger, op. cit. p. 76.

18. Hein, op. cit., p. 551.

19. Ibid., p. 552.

20. Ibid., p. 552.

21. See *Wiltshire Archaeology and Natural History Society,* Vol. 130 (2010).

22. Piggott, Stuart, *Excavation of the West Kennet Long Barrow, 1955-56* (London: Ministry of Works Archaeological Reports No.4, HMSO, 1962).

III - THE CYCLICAL GODDESS

• The Cailleach

1. Davies, J.C., *Folklore of West and Mid-Wales* (Aberystwyth: Welsh Gazette, 1911), p.81.

2. Gundert, Hermann, *Malayalam-English Dictionary*, Ed.

3. *1872* (Kerala, India: Sahythia Pravarthaka Sahakarana Sangham, 2000).

• The Bassingham Plough-Jags

1. Anderson, E.R., The Seasons of the Old English, *Anglo-Saxon England*, 26 (1990, pp. 232-246.

2. Baskerville, C.R., The Bassingham Plough Play, *Modern Philology*, XXI (1924) pp. 241-246.

3. Baskerville, C.R., Mummers' Wooing Plays in England, *Modern Philology*, XXI (1924) pp. 225-240.

4. Brody, A., *The English Mummers and their Plays*: Traces of Ancient Mystery (London: Routledge Kegan and Paul, 1969).

5. Cawte, E., and Helm, A., *English Ritual Drama: A Geographical Index* (London: The Folk-Lore Society, 1967).

6. Chambers, E.K., *The English Folk-Play* (Oxford: 1933).

7. Ellis Davidson, Hilda R., *Scandinavian Mythology* (London: Random House, 1969).

8. Ellis-Davidson, Hilda R., The Sword at the Wedding, *Folklore*, LXXI (1960) pp. 12-28.

9. Gelling, P., and Ellis-Davidson, H.R., *The Chariot of the Sun*, 1965.

10. Grattan, J.H.C., and Singer, C., *Anglo-Saxon Magic and Medicine, Illustrated Specially*

from the Semi-Pagan Text 'Lacnunga', Publications of the Wellcome Historical Medical Museum, new series 3 (London: Oxford University Press, 1952).

11. *The Alkborough Plough Play*, Local History Pamphlet, No.6, (1936), Lincolnshire Library.

12. Balkan Comparisons.

13. Dawkins, R.M., The Modern Carnival in Thrace, *Journal of Hellenic Studies*, XXVI (1906).

14. Wace, A.J.B., North Greek Festivals, *British School of Athens*, XV1, (1909-10).

15. Helm, Alex, *English Mummers' Plays*, Mistletoe Series (Cambridge, D.S. Brewer, 1981) pp. 40-56.

16. Storms, Godfrid, *Anglo-Saxon Magic* (Dordrecht: Springer Science and Business Media, 1948) p. 177.

17. Forbes, Bronwen, *Make Merry in Step and Song: A Seasonal Treasury of Music, Mummer's Plays and Celebrations in the English Folk Tradition* (Woodbury, USA: Llewellyn Publications, 2009) pp. 18-21.

18. *Stamford Mercury* on 8th January, 1847.

• Gaea at the Godmanchester Temple

1. McAvoy, F., The Development of a Neolithic Monument Complex at Godmanchester, Cambridgeshire in Dawson M., Prehistoric Landscapes of the Great Ouse Valley, *CBA Report 119* (2000) pp. 51-6.

2. Keys, David, Godmanchester's Temple of the Sun, *New Scientist*, Volume 129 (29th March, 1991) pp. 29-31.

3. Humble, J. (no date) unpublished report, on Godmanchester, site 432. An assessment of the possible astronomical significance of the post-hole alignments, in *British History Online*, Monuments in Huntingdonshire, pp. xxv-xxiv.

4. Jones, A. (ed.) Godmanchester, Birmingham University Field Archaeology Unit, Monograph Series 6, *BAR*, vol. 364 (2003).

5. Fox, C., *The Archaeology of the Cambridge Region: a topographical study of the Bronze, Early Iron, Roman and Anglo-Saxon Ages, with an Introductory Note on the Neolithic Age* (Cambridge: Cambridge University, 1923).

6. Snow, C. P., *The Two Cultures and the Scientific Revolution*, BBC Rede Lecture (1959). Gaia, as the great creative principal, in *The New Larrouse Encyclopedia of Mythology*, 1959, p.63. and Smith, William (ed.) *Dictionary of Greek and Roman Biology and Mythology*, vol. ii, (London: John Murray Publishers, 1849), p.p. 195-6.

7. Aeschylus, The Persians: The Seven Against Thebes. Word fragment 44, spoken by Aphrodite, and written about c. 520 B.C. From Grene, D. and Lattimore, R. (eds) *Complete Greek Tragedies* (Chicago, USA: University of Chicago, 1953).

8. Rivet, A.L.F. and Smith, C., *The Place-Names of Roman Britain* (Batsford, 1979) p. 364.

9. Ibid., p. 365.

10. Frend, W.C.H., 'The Roman Farm Settlement at Godmanchester', *Proceedings of the Cambridge Antiquarian Society*, Vol. 68 (1977-8).

11. Graves, Robert and Riding, Laura, *Essays from 'Epilogue', 1935-1937 (Lives and letters: the Millennium Graves)* (Manchester: Carcanet Press, 1991).

12. Also in her editorship of Epilogue, by Freidman E., and Clark, A.J. Manchester, 1994, pp. 40-41.

13. Richardson, Dorothy, 'The Essential Egoist', a section within her essay 'Women and the Future', in *Vanity Fair* (April, 1924).

14. Marsden, Dora (ed.) *The Egoist: An Individualist Review*, July (1919) p. 34.

319

15. Heidegger, Martin, 'Building, Dwelling, Thinking,' in *Poetry, Language, Thought* (New York: Harper and Row, 1971) pp. 149-151.

• **The Deepdale Effigy**

1. Barnes, M., *Boston Spa Community Archaeology Report: The Deepdale Effigies* (2007), p. 29.

2. Ibid., p.34.

See also A Prehistoric Portal to the Underworld, *Current Archaeology*, Vol. 206, (November 2006), pp. 33-5.

IV: TOPOGRAPHICAL GODDESSES

• **Spring Heads**

1. *Shorter Oxford English Dictionary* (Oxford: Oxford University Press, 2007).

2. Rowan, *The Folklore of British Holy Wells* (*White Dragon*, August 1996).

3. Ibid., p. 47.

4. Jones, Francis, *The Holy Wells of Wales* (Cardiff: University of Wales Press, 1954).

5. Quiller-Couch, Mabel and Lillian, *Ancient and Holy Wells of Cornwall* (London: Chas J. Clark, 1894).

6. Bord, Janet, *Holy Wells in Britain: A Guide* (Loughborough: Heart of Albion Press, 2008).

7. Simpson, J. and Roud, S., *Oxford Dictionary of English Folklore* (Oxford: Oxford University Press, 2000) pp. 385-6.

8. Branston, Brian, *Gods of the North* (London: Thames and Hudson, 1980) pp. 76-83.

9. Ibid., pp. 26-83.

10. Branston, pp. 147-151.

11. Ibid, pp. 63-4.

12. Guirand, Felix (ed.) *New Larousse Encyclopedia of Mythology* (Paul Hamlyn, 1970) p. 325 and p. 374.

13. Branston, op. cit., pp. 76-82.

14. Ibid., p. 76.

15. Mead from the udders of the goat Heidrun, in Branston, op. cit. p. 94.

16. Branston, op. cit. pp. 208-9.

17. Dames, Michael, *Silbury: Resolving the Enigma* (Stroud: The History Press, 2010) p. 170.

18. *New Larousse Encyclopedia of Mythology*, pp. 61-63.

19. Dames, Michael, pp. 38-9.

20. Dames, Michael, *The Silbury Treasure: The Great Goddess Rediscovered* (London: Thames and Hudson, 1978) p. 97, and p. 114.

21. See Leary, Jim and Field, David, *The Story of Silbury Hill*, (English Heritage, 2010).

22. 'Tissington Well Dressing' in Buckley, Derek Haslam and Williams, Thomas, *A Short History of Tissington and Its Parish Church*, 1965.

23. Southern, Richard W., *Saint Anselm: Portrait in a Landscape* (Cambridge: Cambridge University Press, 1992).

24. Naylor, Peter J., *Ancient Wells and Springs of Derbyshire* (Cromford: Scarthin Books, 1983).

25. Pilsley well dressing, recounted by Mr Ray Bradshaw.

26. Anthony, Wayne and Savage, Jude, *Haunted Derby: Myths, Magic and Folklore* (Derby: Breedon Books, 2007).

27. Ibid., p. 143.

28. Briggs, Katharine Mary, *An Encyclopedia of Fairies: Hobgoblins, Brownies, Bogies and Other Supernatural Creatures* (Pantheon Fairytale and Folklore Library, 1976) pp. 22-23.

29. Campbell, Brian, *Rivers and the Power of Ancient Rome: Studies in the History of Greece and Rome* (Chapel Hill: The University of North Carolina Press, 2012) p. 134.

30. Hope, Robert Charles, *The Legendary Lore of the Holy Wells of England: Including Rivers, Lakes, Fountains and Springs* (London: Elliot Stock Publisher, 1893).

31. – 35. Quiller-Couch, Ancient and Holy Wells of Cornwall.

36. Dames Michael, Silbury: Resolving the Enigma, p. 143.

37. Jones, p. 194.

38. Ibid., p. 162.

39. Leary, J. and Field, D., *The Story of Silbury Hill* (English Heritage, 2010) pp. 140-143.

• River Goddesses

1. *The Ruin of Britain*, c. 600 A.D., (ed) Winterbourne, M., 198 Winterbottom, M. (ed.) *Gildas: The Ruin of Britain and other works, History from the Sources vol. 7* (Chichester: Phillimore & Co., 1978) p. 2.

2. Ekwall, Eilert, *English River Names* (Oxford: OUP, 1968).

3. Milton, John, *Comus* (London: 1637) lines 824-901.

4. Manning-Sanders, Ruth, *The River Dart* (London: Westaway Books, 1951).

5. See Gomme, G.L., on Anthropology and Folklore, *Folklore*, Vol 27 (1916) pp. 408-416.

• A Bevy of 'Silburies'

1. See Alexander Keiller Museum, Avebury.

2. Aldhouse-Green, S. and Pettitt, P., Paviland Cave: contextualizing the 'Red Lady', *Antiquity*, 72 (1998) p. 767.

3. Ibid., p. 766.

4. *Introduction to The Physicians of Myddvai 1*, by Mr Rees, of Tonn (Llandovery: Welsh Manuscript Society, 1861).

5. Rhys, John, The Lady of Llyn y Van Vach, *Y Cymmrodor*, Vol. 4 (1881) p. 147.

6. W.J.W, in *Archaeologia Cambrensis* (Welshpool: 1858), p. 160.

7. Rhys, J., op. cit., p. 177.

8. Sykes, Wirt, *British Goblins:* Welsh Folk-lore, Fairy Mythology, Legends and Traditions (London: 1880).

9. Rhys, John, *Celtic Folklore, Welsh and Manx*, Vol. 1 (Oxford: The Clarendon Press, 1901) p. 29.10. Jones, Gwyn and Jones, Thomas (transl.) *The Mabinogion* (London: Weidenfeld & Nicolson (Everyman), 1993) pp. 1-20, and pp. 148-149.11. Bromwich, Rachel (ed.) *Trioedd Ynys Prydien: The Triads of the Island of Britain* (Cardiff: University of Wales Press, 2006) p. 458.

12. Dames Michael, *Mythic Ireland* (London: Thames and Hudson, 1996) pp. 214, 223 and 232.

13. Bromwich R., op. cit. p. 458.

14. Bartrum, Peter Clement, *A Welsh Classical Dictionary: People in History and Legend Up to About A.D. 1000* (Aberystwyth: The National Library of Wales, 1993) p. 90.

15. Ibid., p. 431-2.

16. Bromwich, R., p. 458.

17. Ibid., p. 459.

18. Pratt, Samuel Jackson, *Gleanings Through Wales, Holland and Westphalia, with Views of Peace and War at Home and Abroad. To Which is Added Humanity; or the Rights of Nature.*

A Poem Revised and Corrected (London: T.N. Longman and L.B. Seeley, 1795).

19. Rhys, J., Welsh Fairy Tales, in *Y Cymmrodor*, No. 6 (1883) pp. 168-178.

20. Pennant, Thomas, *Tours in Wales*, 1773, republished in Rhys, John (ed.) Tours In Wales, Vol. 2 (1883) pp. 325-6.

21. Baring Gould, Sabine and Fisher, John, *The Lives of the British Saints: The Saints of Wales and Cornwall and Such Irish Saints as Have dedications in Britain*, Vol. 1 (London: The Honorable Society of Cymmrodorian, 1907-13) p. 164.

22. Conversation with Frank Cowdray, of All Cannings, June 1973.

23. Coombs, D.G., Mam Tor, In Harding, Dennis W. (ed.) *Hillforts: Later Prehistoric Earthworks in Britain and Ireland* (London: Academic Press, 1976) p. 147.

24. Camden, William, *Britannia: or a Chorographical Description of Great Britain and Ireland, Together with the Adjacent Islands* (London: Mary Matthews for Awnsham Churchill, 1722).

25. Branson, Brian, *Gods of the North* (London: Thames and Hudson, 1980).

26. Alcock, Leslie, in *Proceedings of the Prehistoric Society*, No. 16 (1950) p. 81.

27. Ralph, G. H., The Wrekin, in *The Shropshire Magazine* (September, 1965) p. 19.

28. Burne, Charlotte Sophia, *Shropshire Folklore: A Sheaf of Gleanings* (London: Trubner & Co., 1883) p. 197.

29. Ibid., p. 434

30. *Bye-Gones: Relating to Wales and the Border Counties*, Vol. 1 (Caxton Press, 1871) p. 110.

31. *Bye-Gones*, Feb. 1873 p. 141.

32. Rhys, John, All Around the Wrekin, in *Y Cymmrodor*, no. 21 (1908) pp. 11-60.

33. Ibid., p. 60.

34. Burne, C., op. cit., p. 130.

35. Ibid., pp. 330-332.

36. Danielou, Alain, *Hindu Polytheism* (Princeton: Bollingen Foundation, 1964) p. 226.

37. Aldhouse-Green, op. cit., p. 766-7.

38. Partridge, J.B., Roseberry Topping, *Folklore*, No. 23 (1912) pp. 332-48.

39. Allies, Jabez, *The British, Roman and Saxon Antiquities and Folklore of Worcestershire* (London: J.R. Smith, 1856) pp. 78-81 and pp. 363-5.

40. Derham, Dr William, *Physico and Astro Theology or a Demonstration of The Being and Attributes of God* (London: W. and J. Innys, 1713) p. 70.

41. Palmer, Roy, *The Folklore of Worcestershire* (Hereford: Logaston Press, 2005) p. 296.

42. Ibid.

43. Partridge, op. cit., p. 342.

44. Palmer, op. cit., p. 6.

45. Grose, Francis, *A Provincial Glossary: With a Collection of Local Proverbs, and Popular Superstitions* (London: S. Hooper, 1790) p. 54.

46. Turton, R.B., Roseberry Topping, in *Yorkshire Archaeological Journal* No. 22 (1913) pp. 40-48.

47. Nichols, J.D. (ed.) *Topographer and Genealogist*, Vol. 2 (1853) pp. 405-30.

48. Gutch, E., County Folklore: Examples of Folklore Concerning the North Riding of Yorkshire, North Riding and the Ainsty, *Publications of the Folklore Society* XLV (London: David Nutt, 1901) p. 31.

49. Anon., *Account of Guisborough*, c. 1640, BM MS Cotton, Julius, Evi. At Fol. 431c.

50. Longworth, Ian H., *Regional Archaeology; Yorkshire* (Tring: Shire Publications, 1985) p. 47.

51. Allies J., op. cit., pp. 216-20.

50. Spackman, F.T., The Cruckbarrow Gathering, *Worcester Naturalist Club*, No. 4 (1909-10).

51. Corbett, E.H., Worcester Folklore, *Transactions of the Worcestershire Archaeological Society* (1944) p. 25.

52. Mawer, A., and Stenton F.M., *Place-names of Worcestershire* (Cambridge: 1927) p. 274.

53. Timmings, W., *The History and Antiquities of St Kenelms, or Callums: and Kelmestowe, Salop, the Murder and Burial of Kenelm, King of Mercia* (Stourbridge: 1839) pp. 26-7.

54. Ibid., p. 27.

55. Amphlett, John, *A Short History of Clent* (London: Parker and Co., 1890) p. 26.

56. A Latin Hymn, in *Bodleian. MS 285,* Fol. 80-3.

57. Richard of Cirencester, d. 1402, *Speculum Historiale,* Vol. 1, ii, p. 67.

58. Wilde, Jane Francesca Agnes, *Ancient Legends, Mystic Charms and Superstitions of Ireland* (London: Ward and Downey, 1888) pp. 102-3.

59. Richard of Cirencester, op. cit., p. 33.

60. Johnson, Samuel, *Lives of the Poets,Vol. 11 (*1903) pp. 38-9.

61. Pevsner, Nikolaus, *Worcestershire* (London: Penguin, 1968) p. 178.

62. Farmer, David Hugh, *Oxford Dictionary of Saints* (Oxford: Oxford University Press, 1978) p. 77.

63. Hole, Christina, *English Shrines and Sanctuaries* (London: Batsford, 1954) p. 160.

64. Watkins A., Glastonbury Legends, in Carley, James P. (ed.) *Glastonbury Abbey and the Arthurian tradition* (Cambridge: D.S. Brewer, 2001) pp. 17-18.

65. Dyer, J., *The Penguin Guide to Prehistoric England and Wales* (London: Penguin Books, 1981) pp. 255-6.

66. Williams, A., Clegyr Boia, St David's (Pembrokeshire): excavation in 1943. *Archaeologia Cambrensis*, 102 (1952) pp. 20-47.

67. Dames, Michael, Mythic Ireland (London: Thames and Hudson, 1992) pp. 73-112.

68. Smyth, Daragh, A Guide to Irish Mythology (Newbridge, Ireland: Irish Academic Press, 1998) p. 153.

69. Baring-Gould, S., The Exploration of Clegyr Voia, *Archaeologia Cambrensis*, 3 (1903) pp. 1-4.

70. Dyer, op. cit., p. 282.

71. Dames, Michael, *Merlin and Wales: A Magician's Landscape* (London: Thames and Hudson, 2002) p. 127.

72. Gildas, *The Ruin of Britain and Other Works*, Winterbottom, M. (transl.) (Phillimore & Company, 1978) Section 2, paragraph 4.

73. Rhys, John, in *Hibbert Lectures*, Vol. 2 (1886) p. 421.

74. In Newport Museum, Gwent.

75. Dames, Michael, *The Avebury Cycle* (London: Thames and Hudson, 1976) pp. 145-6.

76. Jones and Jones, op. cit., pp. 66-7.

77. Burnham, Helen, *Clwyd and Powys: A Guide to Ancient and Historic Wales* (Welsh Historic Monuments, HMSO, 1995) pp. 11-12.

78. Gruffydd, W.J., Donwy, in *Bulletin Board of Celtic Studies*, Vol. 7 (1933) pp. 1-4.

79. Jackson, R.H., The Gop, or y Gopa tumulus, Flintshire, in *Archaeologia Cambrensis* 4 (1858) p. 152.

80. Boyd Dawkins, WIlliam, *The Gop,* in *Archaeologia Cambrensis* (1921) pp. 163-171. Boyd Dawkins, William, On the cairn and sepulchral cave at Gop, near Prestatyn, *Archaeologia Cambrensis* 2 (1902) pp. 161-85.

81. Boyd Dawkins, William, Ruthin meeting report, *Archaeologia Cambrensis*, 76 (1921) p. 413.

82. Greenwell, W., Willie Howe in *Archaeologia*, Vol. 52 (1890) pp. 22-4.

83. Ibid., p. 9.

84. Ibid., p. 24.

85. Anderson, J., in *Proceedings of the Society of Antiquaries of Scotland*, No. 1 (1792) pp. 126-142.

86. Drewett, P., The Excavation of Four Round Barrows of the Second Millennium B.C. at West Heath, Harting, 1973-5. *In Sussex Archaeology Collections* Vol.114 (1976) pp. 126-150.

87. MacNeill, Máire, *Festival of Lughnasa: A Study of the Survival of the Celtic Festival of the Beginning of Harvest* (Oxford: Oxford University Press, 1962) p. 357.

88. Dames, Michael, *Mythic Ireland*, pp. 66-9, and pp. 146-9. MacNeill, op. cit., pp. 143-4.

89. Davies, J.A., Roman Coins from Lowbury Hill, in *Oxoniensia*, Vol. 50 (1985) p. 8.

90. Fulford M.G. and Rippon S.J., Lowbury Hill, Oxon: a Reassessment of the Probable Romano-Celtic Temple and the Anglo-Saxon Barrow, *Archaeology Journal of the Royal Archaeological Institute*, No. 151 (1994) pp. 158-211.

91. Cannon, Paul and Henig, Martin, A Sceptre-head for the Matres Cult and Other Objects from West Berkshire, *Britannia*, No. 31 (November, 2000) pp. 358-60.

92. Jencks, Charles, in *The Telegraph*, 1st July, 2009.

93. Dowdall Mark, in *The Times*, 12th September, 2009.

• 'Wonder' of the Peak

1. Hobbes, *Thomas, The Wonders of the Peake in Derbyshire: Translated by a Gentleman of Quality* (London, 1636) p. 40.

2. Cameron, Kenneth and Smith, A.H. (eds.) The Place-Names of Derbyshire, Part 1, Introduction, River-Names etc. High Peak Hundred, *English Place Name Society*, Volume 27, p. 160.

3. Ross, Anne, *Pagan Celtic Britain*: Studies in Iconography and Tradition (Chicago: Academy Chicago Publishers, 1996).

4. Quoted in Camden, William, *Britannia* (London, 1586).

5. Foucault, Michel, *Madness and Civilisation: A History of Insanity in the Age of Reason* (transl.) Howard, R. (New York: The New American Library, 1967).

6. Marcuse, Herbert, *Eros and Civilisation: A Philosophical Enquiry into Freud* (London: Routledge, 1967).

7. Lloyd J. and King E., in *Philosophical Transactions*, Vol. xvi (1771) pp. 250-266.

8. *Place Names of Derbyshire,* op. cit., p. 54.

9. Eve shared an apple with Satan.

10. Dames, Michael, *Merlin and Wales: A Magician's Landscape* (London: Thames and Hudson, 2002) pp. 95-99.

11. Cotton, Charles, *The Seven Wonders of the Peake* (London: J. Wallis, 1683), pp. 30-33.

12. Ibid., p. 34.

13. Leigh, Dr., Beauties of England, quoted by Ainsworth, W. Francis, in *A Descent into Eldon Hole*, in the Peak of Derbyshire (London: L. Davis and C. Reymers, 1700) p. 262.

14. Ibid., p. 261.

15. Raine, Kathleen, *Defending Ancient Springs* (London, Lindisfarne Books, 1967) p. 105.

16. Fiennes, Celia, *The Northern Journey and the Tour of Kent, 1685-1712* (1697) pp. 30-40.

17. Lloyd and King, op. cit., p. 251.

18. Hobbes, op. cit., p. 57.

19. Lloyd and King, op. cit., fig v, p. 265.

20. Cotton, op. cit. p. 33.

21. Purchas, Samuel, *Pilgrim or Microcosmus, or the Historie of Man: Relating the Wonders of his*

Generation, Vanities in his Degeneration, Necessities of his Regenerations (London: 1619) pp. 25-6.

22. Raleigh, Sir Walter, *History of the World* (London: Walter Burre, 1614).

23. Drayton, Michael, *Poly-Olbion* (London: 1619-23) fig. 16.

24. Blake, William, *Jerusalem, The Emanation of the Giant Albion* (London: 1808) p. 66.

25. National Trust, *Guide To the Peak District* (1962)

26. Porteous, Crichton, *Derbyshire* (London: Hale, 1950).

27. Ibid., p. 63.

28. Lloyd and King, op. cit., p. 250.

29. *Folklore, Myths and Legends of Britain* (London: Automobile Association, 1968) p. 288.

30. Ibid., p. 288.

31. Ibid., p. 288.

32. Place names of Derbyshire, Volume 1, op. cit., p. 63.

33. Coombs, D.G., Mam Tor Excavations, *Derbyshire Archaeological Journal*, Vol.99 (1972) pp. 7-51.

34. Ford, T.D. and Rieuwerts, J.H., Odin Mine, Castleton, Derbyshire, *Bulletin of the Peak District Mines Historical Society*, Vol. 6 No. 4 (September, 1976) pp. 7-31.

V. THE CLASSICAL LEGACY

• Roman Silbury

1. Payne, A., Linford, N., Linford, P.K., and Martin, L., Silbury Hill Environs, Avebury, Wiltshire, *English Heritage Report on Geophysical Survey*, (2005) pp. 1-7.

2. Corney, Mark, New Evidence for the Romano-British Settlement by Silbury Hill, *Wiltshire Archaeological and Natural History Magazine* Volume 90 (1997) pp. 139-150.

3. Bewley, R., reported in *The Guardian*, July 10th, 2007.

4. Wilkinson, J., A Report of Diggings Made in Silbury Hill, and In the Ground Adjoining, *WAM (Wiltshire Archaeology Magazine)*, Volume 11 (1869) p. 116.

5. Bewley, R., quoted by Kennedy, M., in *The Guardian*, March 10th, 2007.

6. Ibid.

7. Ibid.

8. Neale D.S. in *Britannia*, Vol. 37 (2007) pp. 72-3.

9. Robinson, P., *Religion in Roman Wiltshire*, 1955, pp. 160-2. Later published in Ellis, P. (ed.) *Proceedings of the Eighteenth Annual Theoretical Roman Archaeology Conference* (2009). Roman Wiltshire and After, *Wiltshire Archaeological and Natural History Society*, 147-164, pp. 5–38. Cunliffe, B., Robinson, P. (2001).

10. De la Bédoyère, G., *Guy, Roman Towns in Britain* (London: B.T. Batsford, 1992) p. 103.

11. Perowne S., *Roman Mythology* (London: Paul Hamlyn, 1969) pp. 52-3.

12. Dames, Michael, *The Silbury Treasure* (London: Thames and Hudson, 1976).

13. Ovid, *Metamorphoses*, (transl.) Innes, Mary M. (London: Penguin, 1955) p. 136.

14. Ibid., p. 137.

15. Ibid., p. 138

16. Ibid., p. 140.

17. Field, David, The Investigation and Analytical Survey of Silbury Hill, *English Heritage Archaeological Investigation Report Series* AI/22/2002 (Swindon: English Heritage, 2002) p. 43.

18. Ibid., p. 137

19. Ibid., pp. 137-8.

20. *New Larousse Encyclopedia of Mythology* (London: Paul Hamlyn, 1970), pp. 205-7.

21. Ovid, op. cit., p. 139.

22. Dames, Michael., *The Silbury Treasure*, 1976, and *The Avebury Cycle*, 1977.

23. Richardson, Nicholas J. (ed.) *The Homeric Hymns to Demeter* vol. 2 (Oxford: Oxford University Press, 1974) p. 197 and p. 313.

24. Powell, A.P., *Wessex Archaeological Report*, Vol. 8 (2007) p. 55.

25. Roman Coins From the Silbury Region, *Marlborough College Natural History Society*, Vol. 37 (1869) p. 41.

26. Dames, Michael, *The Avebury Cycle* (1996) pp. 219-222.

• Recovering Britain's Classical goddesses

1. Dames, Michael, *The Avebury Cycle* (London: Thames and Hudson, 1977).

2. Cunliffe, B. *Roman Bath Discovered* (Stroud: Tempus Publishing, 2000).

3. Guirand, Felix, *New Larousse Encyclopedia of Mythology* (London: Paul Hamlyn, 1970) p. 207.

4. Ibid., p. 207.

5. Cunliffe, B., op. cit., p. 21.

6. Hadrian, copper coin of A.D. 119. Britannia is shown bareheaded. Other Britannia coins of A.D. 154 and 290; and Britannia is depicted on a coin from the Charles II period, 1672.

7. Collingwood, R.G. and Wright R.P., *The Roman Inscriptions of Britain, I, Inscriptions on Stone* (Oxford: Clarendon Press 1965), no. 2175.

8. Ibid., no. 643.

9. Ibid., no. 51, p. 123.

10. A belief still flourishing in the 17th Century. See Drayton M., *Poly-Olbion* (1612-24).

11. Julia Domna, wife of Emperor Septimius Severus. See Collingwood & Wright, op. cit., no.1791 found at Carvonan.

12. Ibid., no. 910. To Julia Mammea.

13. Ibid., over 30 instances of Matres dedications.

14. Ibid. See also no. 2050, an altar 'to his own Mother Goddesses of the household', set up by Asinios Senilis.

15. Ibid., Colchester, no. 192.

16. Ibid., Carlisle, no. 951.

17. Ibid., no. 2121, on an altar, 12 in by 26 in (30 cm by 66 cm).

18. Ibid., no. 324, 'To Salus the Queen'.

19. Worshipped in Rome from 311 B.C. onwards.

20. Collingwood and Wright, op. cit., no. 316, a dedicatory tablet, from Caerleon. And no. 318, combining Fortuna with Bonus Eventus.

21. Ibid., no. 968.

22. Temple of Isis, in Southwark, on the South Bank of the Thames.

23. See Toynbee, Jocelyn M.C., *Art in Britain Under The Romans* (Oxford: Oxford University Press, 1964), p. 65.

24. Diana's temple at Caerleon, restored by Flavius Postimius, mid-3rd Century.

25. Collingwood and Wright, op. cit., no 140, at Bath.

26. An altar to Diana, and 1209, another 'altar to the goddess Diana', set up by Alio Timo.

27. Ibid., Chesters, No. 1448. Celebrated on May 1st.

28. See Wheeler, Sir Mortimer, *Temple of Nodens Excavation Report* (Oxford, 1932) p. 68 and plate 24.

29. Collingwood and Wright, op. cit., no. 886, Old Carlisle. Aeternae, alias Roma.

30. Ibid., no. 1124, inscription in Greek at Corbridge, Northumberland.

31. See also, Collingwood and Wright, op. cit., no 2065, her altar set up by Apollonius the priest. And no. 1275, from Caerleon.

32. Thetys, Ibid., no. 663.

33. Ibid., no. 813, 'to Juno, Diana Hermione set this up'.

34. Latin *tacita*, 'silent'. See her 'finger on lips' figurine, found in Worcester.

35. Ibid., No. 609, Overborough, Co. Durham.

36. Ibid., Collingwood and Wright, no. 1777 at Carvoran.

37. Also referred to as Victoria, on no. 628, from Castleford, West Yorkshire, and as Caelestis, 'heavenly', at Corbridge. She is a regional native deity invested with many Classical attributes.

38. Ibid., Collingwood and Wright, no. 460, Fontes et Nymphis, Chester. And at Greta Bridge, an altar by Brica to her daughter Januaria, no. 744.

39. Clarke, G., Rigby, V., and Shepherd, J.D., The Roman Villa at Woodchester, *Britannia*, Vol. 13 (1982) pp. 197-228.

40. The Rudston mosaic is now in Hull Museum.

41. The Goddess Arnemetiae presided over Buxton's holy spring, later renamed St Anne's well.

42. Collingwood and Wright, op. cit., no 926, 'To all the deities'; and no. 811, Maryport, 'To all the gods and goddesses'.

43. White paste Venus figurines, imported from Gaul.

44. Collingwood and Wright, op. cit., Garmangabis, at Lanchester, no. 1074.

45. Ibid., no. 890, Bellona, sister of Mars.

46. Pax (Peace) Ibid., no 1791.

47. Ibid., no. 1897. And no. 2043, from Burgh by Sands, Cumbria.

48. Ibid., no 1135, 'with good fortune to the goddess Panthea'.

49. Ibid., no. 1454, 'to the goddess Ratis'.

50. Ibid., Benwell, Northumbria, no. 1331, 'To the Three Witches' (*Lamiis Tribus.*)

51. Cosh, Stephen R. and Neal, David S., *The Roman Mosaics of Britain*, Vol. 1 (Aldborough: Society of Antiquaries, 2002), pp. 123-25.

52. Collingwood and Wright, op. cit., no. 841.

53. Ibid., no. 2096, 'Sacred to the Goddess Harimella'.

54. Ibid., no. 1780. *See also Archaeologia Aeliana*, (Journal of the Society of Antiquaries of Newcastle Upon Tyne), Vol. xxi, (1943), p.203.

55. Collingwood and Wright, op. cit., no. 1207.

56. Ibid., no. 2108.

57. As around the famous example in Cirencester.

58. At Bignor, West Sussex.

59. At Littlecote, Berkshire.

60. Crowned female deities, whose role was to protect the wellbeing of the city, placed under their care.

• **Written in Stone**

1. *The Antonine Itinerary*, the Iter Britanniarum section. 3rd Century A.D. Author unknown.

2. Sulis, see O.Ir. *sùil* eye; cognate with proto-Celtic Suli – sun. see *Dictionary of the Irish Language* (Dublin: Royal Irish Academy, 1976).

3. Dames, Michael, *Mythic Ireland* (London: Thames and Hudson, 1992) pp. 62-4, and pp. 108-9.

4. Cunliffe B. (ed.) *The Temple of Sulis Minerva at Bath, Vol. II, Finds from the Sacred Spring*, Oxford University Committee for Archaeology, Monograph 16 (Oxford: OUP, 1988).

5. Cunliffe, Barry W., *The Roman Baths, A View Over 2000 years: The Official Guide* (Bath: Bath Archaeological Trust, 1993).

6. Green, Miranda J., *The Gods of Roman Britain* (Oxford: Osprey Publishing, 1983).

7. Cunliffe, B., *The Temple of Sulis Minerva at Bath* 1988.

8. Tomlin, Roger S.O., Curses from Bath, in Cunliffe, B., op. cit., 1988.

9. Clayton, J, Description of Roman Remains Discovered near to Procolitia. in *Archaeologia Aeliana*, Volume 8, (1880) pp. 6-21.

10. Ibid., pp. 7-8.

11. See watercolour by Mossman, F., 1878, now in Tullie House Museum and Art Gallery, Carlisle, and Coventina's Temple and Well at Procolita, *Illustrated London News*, November 15th, (1876).

12. Dinneen, P.S., *An Irish-English Dictionary* (Dublin: Gill and Son, Irish Texts Society, 1904).

13. Allason-Jones, L. and McKay, B., *Coventina's Well; A Shrine on Hadrian's Wall* (Chollerford: Chesters Museum, 1985) p. 12 and p. 48.

14. Carrawbugh, in Collingwood, R.G. and Wright R.P., *The Roman Inscriptions of Britain, I, Inscriptions on Stone* (Oxford: Clarendon Press 1965) pp. 486-489.

15. Allason-Jones, L. and McKay B., op. cit. p. 12. An Imperial edict of A.D. 391 ordered that all pagan temples were to be closed down.

16. The Matres, see Green, Miranda J., *The Gods of Roman Britain* (Princes Risborough: Shire Publications, 2003).

17. Haverfield, F., The Mother Goddesses in *Archaeologia Aeliana*, 2nd Series (1892) pp. 314-335.

18. Collingwood and Wright, op. cit., p. 301.

• **An Effigy of Sabrina**
1. Drayton, Michael, *Poly-Olbion* (London: c. 1612-1620).

2. Tacitus, *Annals*, 1st Century A.D.

3. Geoffrey of Monmouth, *The History of the Kings of Britain*, c. 1130, translated by Thorpe, Lewis (London: Penguin Books, 1966).

4. Milton, John, *Comus*, (1637) lines 824-902.

5. Professor Lapworth, 1890, Birmingham University.

VI - HORSE GODDESSES

• **The Uffington Horse Deity**
1. Hughes, Thomas, *The Scouring of the White Horse* (Cambridge: Macmillan and Co., 1859).

2. Woolner, Diane, New Light on the White Horse, *Folklore*, 78 (1967) p. 99.

3. Guirand, Felix (ed.) *Larousse Encyclopedia of Mythology* (London, Hamlyn, 1968) p. 163.

4. Dames, Michael, *Mythic Ireland* (London, Thames and Hudson, 1992), pp. 60-66.

5. Corinium Museum, Cirencester.

6. Hughes, op. cit., p. 10.

7. Woolner, op. cit., p. 95.

8. Hughes, op. cit., p. 15.

9. Gelling, Margaret, The Charter Bounds of Aescesbyrig and Ashbury, in *Berkshire Archaeological Journal*, 63, (1967-8), p.6.

10. *Bosworth-Toller Anglo-Saxon Dictionary* (www.bosworthtoller.com).

11. Gelling, op. cit., p. 8.

12. Hughes, op. cit. p. 77.

13. Ibid., p. 107 and pp. 108-9.

14. Jackson's Oxford Journal, 29th May (1780).

15. Gelling, Margaret, *Place-Names of Berkshire*, Vol. III, p. 801 and Ekwall, E., *English River Names* (Oxford: The Clarendon Press, 1928) p. 306-7.

16. *Mythic Ireland*, op. cit., p. 179-80.

17. Piggott, Stuart, The Uffington White Horse, *Antiquity*, V (1931) p.37 and Ross, Anne, *Pagan Celtic Britain: Studies in Iconography and Tradition* (Abingdon: Routledge, Kegan, Paul, 1967) pp. 14-18.

18. Ralf de Deceto, *Wonders of Britain*, c. 1200, quoted by Woolner, op. cit., p. 99.

19. Gelling, Margaret, Place-Names of Berkshire, Part II, *English Place Name Society*, Vol. 50 (Cambridge: Cambridge University Press, 1974) p. 369.

20. Dames, Michael, *The Silbury Treasure: The Great Goddess Rediscovered* (London: Thames and Hudson, 1976).

21. British bronze figurine, British Museum.

22. Jones, Gwynn and Thomas, *The Mabinogion* (London: Everyman's Library edition, 1889 revised edition) p. 17.

23. Helm, Alex, *The English Mummers' Play* (London: Rowman and Littlefield, 1981) p. 56.

• Uisneach Horses Revisited

1. Macalister, R.A.S. and Praeger, R. Lloyd, Report on the Excavation of Uisneach, *Proceedings of the Royal Irish Academy*, Part C, Vol. 39 (1929-31) pp. 54-83.

2. Dames, Michael, *Mythic Ireland* (London: Thames and Hudson, 1992) p. 199.

3. Ibid., pp. 195-7.

4. O'Rahilly, Thomas F., *Early Irish History and Mythology* (Dublin: Dublin Institute for Advanced Studies, 1946) pp. 287-307.

5. Ibid., p. 290.

6. Ibid., pp. 287-89.

7. *Mythic Ireland*, op. cit., p. 186 and p. 218.

8. Ross, Anne, *Pagan Celtic Britain: Studies in Iconography and Tradition* (London: Routledge, 1967), p. 321; and O'Rahilly, *Early Irish History and Mythology* pp. 303-5.

9. Dames, Michael, *The Silbury Treasure* (London: Thames & Hudson, 1976), p. 14, and *Silbury, Resolving The Enigma* (Stroud: The History Press, 2008) p. 101.

10. MacNeill, M., *The Festival of Lughnasa* (Oxford: Oxford University Press,1962) pp. 203-4.

11. O'Riordain, Sean P., *Antiquities of the Irish Countryside* (London: Methuen, 1964) p. 5.

12. *Mythic Ireland*, op. cit., p. 229.

13. Keating, Geoffrey (ed.) The History of Ireland, *Irish Texts Society*, Vol. 2 (London: 1908) p. 249.

14. McCormick, F., The Horse in Early Ireland, *Anthropozoologica*, 2007, Vol. 42, p. 85.

15. *Early Irish History and Mythology*, pp. 286-7, and p. 297.

16. *Revue Celtique*, 16, (1895) (transl. Stokes) p. 60.

17. *Dictionary of the Irish Language* (Dublin: Royal Irish Academy).

18. Lebor Gabála Érenn, Vol. 5, pp. 35-7.

19. Lebor Gabála Érenn, Vol. 5, para. 392, pp. 35-7.

20. Frewin Hill, see *Mythic Ireland*, op. cit., pp. 240-2.

21. Macalister and Praeger, op. cit., p. 74.

22. Lucas, A.T., Souterrains: the Literary Evidence, *Béaloideas,* Vol. 39-41 (1971-73) p. 169.

23. *Mythic Ireland*, op. cit., p. 68-9.

24. Ibid., pp. 106-7.

25. Ibid., p. 155.

26. Ibid., pp. 257-8.

27. Ibid., p. 48, p. 127, p. 210, p. 253.

28. Ibid., p. 123, p. 131.

29. Ibid., p. 200.

30. Ibid., pp. 208-9.

31. Ibid., pp. 154-5.

32. Macalister and Praeger, op. cit, p. 31.

33. Camden, William, *Britannia* (1607, English translation by Holland, Philemon, 1610) p. 301.

34. Macalister and Praeger, op. cit., pp. 61-2.

35. *Jones' Celtic Encyclopedia*, p. 1.

36. Bergin, O., and Best, R.I., Tochmarc Etaine, written c. A.D. 1000., in *Eriu*, Vol. 12 (1938) p. 145, para. 5.

37. Rolleston, T. W., *Myths and Legends of the Celtic Race* (New York: Schocken Books, 1986). First published 1911. p. 136.

38. *Cath Magh Turedh: The Second Battle of Maige Tuired*, Gray, E.A. (ed) Irish Texts Society, LII, Naas (1982) pp. 47-9.

39. Jones' Celtic Encyclopedia, Morrigan; and Cath Magh Turedh, p. 45.

40. Macalister and Praeger, op. cit., p. 74.

41. Macalister and Praeger, op. cit. p. 75.

42. Ibid., p. 75.43. Ibid., p. 75.

44. ibid., p. 70, pp. 75-6.

45. See fig. 1a.

46. Keating, Vol. 2, p. 247.

47. Macalister and Praeger, op. cit., p. 73.

48. Ibid., p. 72.

49. *Mythic Ireland*, op. cit., p. 229.

50. Macalister and Praeger, op. cit., p. 69.

51. Ibid., p. 73.

52. Smyth, Daragh, *A Guide to Irish Mythology* (Dublin: Irish Academic Press, 1988) p. 109

53. Macalister and Praeger, op. cit., p. 73.

54. Dames, Michael, *Merlin and Wales* (London: Thames and Hudson, 2002) pp. 131-2.

55. *Merlin and Wales*, op. cit., p. 154.

56. *Cath Magh Turedh*, pp. 47-9.

57. *The Mabinogion*, pp. 16-19.

58. *Mythic Ireland*, op. cit., pp. 202-4.
59. Macalister and Praeger, op. cit., p. 72.
60. Ibid., p. 72.
61. Ibid., p. 71.
62. Ibid., p. 71.
63. *Mythic Ireland*, p. 79.
64. *Festival of Lughnasa*, op. cit., p. 398.
65. O'Meara, op. cit., p. 110.
66. *Dictionary of the Irish Language.*
67. *Ibid.*
68. Macalister and Praeger, op. cit., p. 82.
69. *Dictionary of the Irish Language, op. cit.*
70. Macalister and Praeger, op. cit., p. 67.
71. *Mythic Ireland*, op. cit., pp. 63-67.
72. *Cath Magh Turedh*, p. 25, paragraph 6.
73. Keating, G., *The History of Ireland*, Vol II (Dublin: Irish Texts Society, 1901-13,) p. 247.
74. Koch, John T., *Celtic Culture: A Historical Encyclopaedia* (Santa Barbara: ABC Clio, 2006) p. 195.
75. Keating, op. cit., Vol. 2, p. 247.
76. Macalister and Praeger, op. cit., p. 93.
77. *Mythic Ireland*, op. cit., p. 215.
78. Ibid., p. 147.
79. Ibid., p. 82.
80. Ibid., p. 58 and p. 82.
81. Ibid., p. 65.
82. *Festival of Lughnasa*, op. cit., p. 10.
83. Ibid., p. 321, and Smyth, op. cit., p. 142.
84. MacNeill, op. cit., pp. 3-10.

• Lady Godiva of Coventry

1. Lancaster, J.C., *Godiva of Coventry* (London: Coventry Corporation, 1967) pp. 44-5.
2. Victoria County History: *Warwickshire*, Volume 8, pp. 4-5.
3. Florence of Worcester.
4. William of Malmesbury.
5. Ellis Davidson, H. R., The Legend of Lady Godiva, in *Folklore*, Vol. 80 (1969) pp. 107-121.
6. Roger of Wendover, *Flores Historiarum.*
7. Matthew of Westminster.
8. Ellis Davidson, op. cit. p. 118.
9. Graves, R. (ed.) *The White Goddess: A Historical Grammar of Poetic Myth* (London: Faber & Faber, 1961) p p. 403-5.
10. Stenton, D.B. *The Place Names of Warwickshire* (Cambridge: Cambridge University Press, 1929).
11. Ekwall, Eilert, *The Concise Oxford Dictionary of English Place Names* (Oxford: OUP/ Clarendon Press, 1960) p. 126.
12. *Yellow Pages* telephone directory, Coventry, *2010.*
13. *Historic Coventry Forum*, River Arthur, Jan. 1970.

14. Clayton, J., *Archaeologia Aelina*, Vol 8 (1880) p. 9.

15. Graves, op. cit., pp. 403-5.

16. Victoria County History: *Warwickshire*, op. cit. pp. 246-7.

17. Dormer-Harris, M. *The Misericords of Coventry*, in Birmingham Archaeological Society, Vol. 52 (1927) pp. 246-60.

18. Palmer, Roy, *Folklore of Warwickshire* (London: B.T. Batsford, 1976) pp. 135-6.

19. Ibid., pp. 137-8.

20. *The Coventry Herald*, 25th April, 1845.

21. *Coventry and Warwick News*, 15th November, 1999.

22. *The True Story of Lady Godiva*, Belgrade Theatre, 1994.

23. Cartwright, L.J., *A few more notes on Southam* (1913). Southam Library.

24. Ibid. The procession survived until 1845.

25. Opie, I., and Opie, P., *The Oxford Dictionary of Nursery Rhymes* (Oxford: Oxford University Press, 1951, 2nd edition, 1997) pp. 65-7. One version has the hobby horse riding to Coventry Cross.

VII. CENTRES AND EDGES

• Some Centre Points

1. Eliade, M., Symbols of the Centre, in *Images and Symbols* (Princeton, Princeton University Press, 1991), pp. 48-51.

2. Quoted in Dumezil, Lyle E., Three Functions and Indo-European Cosmic Structure, *History of Religions*, Vol. 22 (1982) pp. 25-44.

3. Dames, Michael, *Mythic Ireland* (London: Thames & Hudson, 1992) pp. 194-207.

4. *Taliesin's Travels: A Demi-god at Large* (Loughborough: Heart of Albion Press, 2006) pp. 186-191.

5. *Gazetteer of English Place-Names* (http://placenames.org.uk).

6. *The Topographical Collections of John Aubrey for Wiltshire, 1659-70* Jackson, J. E. (ed.) (Devizes: 1862).

7. Thompson, E.P., *Customs in Common* (London: Merlin Press, 1991) pp. 100-102.

8. Ibid.

• Myths of the Lower Severn

1. Rivet, A.L. and Smith, C. *The Place Names of Roman Britain* (London, 1979).

2. Tacitus, (transl.) Hutton, M., Annales, Bk II (London, 1911).

3. O'Rahilly, T.F., *Early Irish History and Mythology* (Dublin, 1944).

4. Macdonell, A.A., *A Practical Sanskrit-English Dictionary* (Oxford: Oxford University Press, 2004).

5. Morris-Jones, Sir J., *An Elementary Welsh Grammar* (London: Clarendon Press, 1921).

6. *Hafren*, in *Liber Landavensis*, 12th Century manuscript (Oxford: Jesus College Library) p. 135.

7. Pughe, W. Owen, *A Dictionary of the Welsh Language* (London: Thomas Gee, 1803).

8. Hooke, D., *The Anglo-Saxon Landscape: The Kingdom of the Hwicce* (Manchester: Manchester University Press, 1985).

9. Dames, Michael, *Mythic Ireland* (London: Thames & Hudson, 1996) pp. 181-2 and pp. 237-9.

10. Ekwall, Eilert, *English River Names* (London: OUP, reprinted 1968) and Ross, Anne, *Pagan Celtic Britain* (London: Routledge and Kegan Paul, 1967).

11. *Culhwch. and Olwen.* transl. Davies, S. (Oxford: 2007) pp. 22-35.

12. Bromwich, R. (ed. and trans.) *Trioedd Ynys Prydein (The Triads of the Isle of Britain)* 3ʳᵈ edit. (Cardiff, 2006).

13. Bartrum, P.C., *A Welsh Classical Dictionary* (Aberystwyth: 1993).

14. Glyn Cuch, a valley that runs along the Pembrokeshire-Carmarthenshire border. See also Langham, A.F., *The Island of Lundy* (Stroud: The History Press, 1994); and *Mabinogion,* (ed.) Davies, op. cit., p. 3 and p. 6.

15. Chanter J.R., *A History of Lundy Island* (Barnstaple, 1871).

16. Drayton, M., *Poly-Olbion* (London, 1612-22).

17. Stowe, John, *The Chronicles of England from Brute to the year of Christ* 1580 (London, 1592).

18. *Preiddeu Annwn* (the Spoils of Annwn) (transl.) Williams, R., in Skene, W.,F., Four *Ancient Books of Wales* (Edinburgh, 1868).

19. Quinnell, H., *Prehistoric and Roman Pottery on Lundy,* in Rep. of the Lundy Field Society, Vol 54 (1992) pp. 89-92.

20. Triad 52, op. cit., Geirioedd.

21. Ibid., Bromwich R., note.

22. Bromwich, R., in *Troiedd Ynys Prydein (TYP)* and *Geiriadur Prifysgol Cymru* (Cardiff, 2016).

23. Coleridge, S.T., *Biographia Literaria,* 2 vols. 1817. Book II, p. 257.

24. Wordsworth, William, Lines Composed Above Tintern Abbey, published in his *Lyrical Ballads* (Bristol, 1798).

25. The Bay of Fundy, Nova Scotia, Canada, claims the highest tidal range.

26. *TYP,* op. cit., p. 78.

27. *Oxford English Dictionary.*

28. *Geiradur Prifysgol Cymru (Dictionary of the Welsh Language)* (Cardiff: University of Wales, 2004).

29. A sea god, whose home was on the Isle of Man. See Dames, Michael, *Mythic Ireland* (London, 1992) p. 269.

30. Leland, J., *Itineraries* (London, 1538-43).31. Defoe, Daniel, *A Tour Through the Whole Island of Britain* (London: 1724).

32. Ellis, F., A bronze figurine from Aust Cliff, in *Transactions of the Bristol and Gloucester Archaeological Society (*TBGAS), Vol. 23 (1900).

33. *British Museum Guide to Iron Age Antiquities,* 2nd ed (1925) p. 148, fig. 173.

34. Smith, S., in *TBGAS,* Vol. 66 (1945) p. 255.

35. *New Larousse Encyclopedia of Mythology* (New York: Hamlyn, 1970) p. 211.

36. Graves, R., *I, Claudius* (London: 1934), and Eckwall, E., *Concise Dictionary of English Place-names* (1936) p. 18, 'Aust is Latin Augusta'.

37. Bede, *Ecclesiastical History of the English People,* Book II, Part 2.

38. Welch, F.B.A., *Geology of the Bristol and Gloucester District* (1935) p. 34-40.

39. Ibid., p. 40.

40. Ovid, *Metamorphoses* (transl.) Innes, M.M., (London: Penguin Books, 1955), pp. 136-145.

41. Langley, Martin and Small, Edwina, *Estuary and River Ferries of South-west England* and Walters (Waine Research Publications, 1984).

42. Peel, J.H.B., *Portrait of the Severn.* 2nd ed. (London: Hale, 1980).

43. Jordan, Christopher, *Severn Enterprise:* The Story of the Old and New Passage Ferries (Ilfracombe: Arthur H. Stockwell, 1977).

44. Rudder, Samuel, *A New History of Gloucestershire* (Gloucestershire: 1779) p. 47.

45. Hatherley, R., Aust, 1650-1720, in *Notes on Bristol History*, No. 5 (1962) p. 85.

46. ibid., p. 15-17.

47. Hutchins O.J. , St Twrog's Chapel, *Severn and Wye Review*, No. 4 (1971) p. 81.

48. Mesolithic Footprints at Gold Cliff, *The Severn Estuary Report Committee* (Lampeter, 1992-1995).

49. The Severn-Wye Bridge, named as the Sri Chimnoy Peace Bridge, on 15th November, 1991.

50. Eck, D.L., in History of Religion, Vol. 20, 1981, p. 323-345. *India's Tirthas.Sacred Crossings in Sacred Geography.*

51. Ballowal barrow, alias Carn Gluze. See Dyer, James, *Southern England: an Archaeological Guide* (London: Faber & Faber, 1973), p. 35. Also Rhys, J., *Celtic Folklore*, Vol II, p. 506.

52. Welsh col. (1) The awn or head of an ear of corn; (2) A bee's sting. *Geiriadur Pryfisgol Cymru.*

53. Bromwich, R., in *TYP*, No. 26 and 26w (Cardiff: 1928). Aber Tarogi joins the Severn at Caldicot Point.

54. But it previously entered the Severn at Sudbrook, says Thomas, R.J., in *Enwau Afonydd a Nentydd Cymru*, (Gwasg Prifysgol Cymru, *1938*).

55. *TYP*, 26w, op. cit., p. 47.

56. Mynydd Gwyn, alias Gray Hill. See Chadwick, A., and Wickstead, H. *in Archaeology in Wales*, Vol. 4 (2002).

57. *TYP*, 26w, op. cit., p. 48

58. Ibid., p. 58.

59. Culhwch and Olwen, *The Mabinogion* (transl.) Davies, Sioned (Oxford: OUP, 2007) p. 212.

60. Ibid., p. 212.

61. Irish *triath*, (1) 'boar; (2) wave'.

62. *The Mabinogion*, op. cit., p. 212.

63. Ibid. p. 212 and note Davies, D., p. 286.

64. Ibid., p. 212.

65. *The Mabinogion*, op. cit., p. 212.

66. Woodward, Horace B., *The Geology of England and Wales*, 2nd ed. (London: George Philip & Son, 1887) p. 241.

67. The bridge was completed in 1966.

68. Dames, Michael, *The Silbury Treasure* (London: Thames & Hudson, 1976) p. 164.

69. *The Mabinogion*, op. cit., p. 212.

70. Nennius, Historia Brittonum: The Wonders of Britain, A.D 829, in Morris, John (ed. & trans.), Nennius: British History and the Welsh Annals (London: Phillimore, 1980).

71. Ibid., pp. 40-41.

72. *GPC*, op. cit.

73. *Dictionary of the Irish Language.*

74. *Nettleship, J., Llyn Llyw, Caerwent Historical Trust Papers* (1995).

75. *Folklore*, August 2008.

76. Barber, Chris, *Exploring Gwent: A Walker's Guide to Gwent, Land of History and Legend* (Clifton: Regional Publications (Bristol), 1984).

77. Nash-Williams, V.E., Sudbrook Fort, *Journal of Antiquities* (1936).

78. *Historia Brittonum*, op. cit., para. 69; and Geoffrey of Monmouth, *History of the*

Kings of Britain, 1136, Wright, Neil (ed.) (London: The Folio Society, 1984) p. 106.

79. A 7th Century poem by Moliant Cadwallon describes 'beautiful Porth Skewett as one of the three chief ports of Wales (see Aldhouse-Green, Miranda, and Howell, Ray (eds) *Gwent County History, Volume 1: Gwent Prehistory and Early History* (Cardiff: University of Wales Press, 2004).

80. *Archaeology in Wales*, Vol. 42, op. cit., p. 102.

81. Children, G., and Nash, G.A., *Prehistoric Sites of Monmouthshire* (Hereford: Logaston Press, 1996).

82. Gwynn Jones, T., *Welsh Folk Lore and Folk Custom* (Cardiff, Methuen, 1929) p. 148.

83. Dames, Michael, *Mythic Ireland*, pp. 247-258.

84. *The Mabinogion*, p. 198.

85. Ibid., p. 203.

86. Ibid., pp. 204-5.

87. Darwin, Charles, *The Origin of Species by means of Natural Selection* (London: John Murray, 1859).

88. Young, Arthur, *A Six Weeks Tour Through the Southern Counties of England and Wales* (London: W. Strahan, W. Nicoll, 1768) pp.132-3.

89. Nodens, in Report on the Excavation of the Prehistoric Roman and Post-Roman Site in Lydney Park, Gloucestershire, by Wheeler, R.E.M. and Wheeler T.V, *Society of Antiquaries* (London: OUP, 1932).

90. Ibid., p. 67.

91. The Broad-Stone, *A History of the County of Gloucester*, Vol 10 (*Gloucestershire Victoria County History, 1972*), p. 981; and Greene, W.H., Jack o' Kent, the Devil and the Pigs, in *Chepstow Weekly Advertiser* (1865).

92. Rudder, S., op. cit., p. 150, and Nicholls, H.G., Nicholls's Forest of Dean: an historical and descriptive account and Iron Making in the Olden Times (Dawlish: David and Charles; Reprint edition, 1966).

93. Paul, J., Floods on the Severn, *TBGAS*, Vol. 78 (1959).

94. Rhonabwy's Dream, in *Mabinogion*, op. cit., p. 2.

95. Nennius, op. cit., p. 40.

96-97. Rowbottom, F.W., *The Severn Bore* (Dawlish: David and Charles, 1964) p. 1.

98. Blagden, Sir C., *Observations on the head of the tide on the river Severn*, in *TBGAS*, Vol. 97 (1979) pp. 123-5.

99. Meas Mawr, alias Mag Mor, 'The Great Plain', was the heartland of the gods, where 'men and maidens lived together without shame, and where music always sounded'. See Matthews, John and Caitlin, *The Aquarian Guide to British and Irish Mythology* (London: Aquarian Press, 1998) p. 113.

100. Modron, the Welsh Mother Goddess, deity of rivers and plains, seen here in both her liquid and solid forms, as the Severn divides briefly to reveal Alney Island.

101. Fiennes, Celia, *Journal*, (London: 1695). 'A low moist place; and therefore one must travel on causeys', p. 197. And Leland J. *Itinerary* vol. II, p. 58. 'In this cawsey be dyvers double-arched bridges to drene the medows at flodes.'

102. Leland, J., op. cit., p. 57 and 63.

103. *Mabinogion*, op. cit., p. 205.

104. Cei is often described as fair and tall.

105. His regular companion, the knight Bedwyr's name, may derive from Welsh bedw, 'birch tree', used in Wales as the Maypole, and as the centre of summer half rejoicing.

106. *Mabinogion*, op. cit., p. 205.

107. Ibid., p. 205.

108. Ibid., p. 270 and p. 178.

109. Bromwich, R., in *Troiedd Ynys Prydein*, note 685, p. 132.

110. Richmond, I.A., and Crawford, O.G.S., *The British Section of the Ravenna Cosmography* (London: Society of Antiquaries, 1949) p. 39.

111. Dames Michael, *The Silbury Treasure* (London, 1976) and *The Avebury Cycle* (1978).

112. Jones, Mary, *Jones's Celtic Encyclopedia* (website).

113. In the 7th Century A.D.

114. *Jones's Celtic Encyclopedia*.

115. *Journal of the Dumfries and Galloway Natural History Society*, Vol. 31, p. 35. The Clacmamanstane is Mabon's.

116. The Roman name, *Glevum*, means 'bright, shining'. See Rivet, A.L.F and Smith, C., *The Place-Names of Roman Britain* (London: Batdford, 1979), op. cit.

117. As in his Severn and Somme poems of 1917: 'Forget me not quite, you Severn meadows'.

118. *The Mabinogion*, op. cit., p. 207.

119. op. cit., p.4, p.p. 6-7..

120. Winterbottom, Michael (ed.) *Gildas: The Ruin of Britain and other works*, (London: Phillimore,1978), p.p.1-3. And *The Mabinogion*, p.p. 4, 6 and 7.

• Brigid's Scottish Home

1. Watson E. C., Highland Mythology, *The Celtic Review* (1908-9) pp. 48-65.

2. Macdonald, Norman, *Gaelic Folklore: Natural Objects With Supernatural Powers* (The Central Register of Folklore Research in Great Britain, 1958).

3. MacLennan, Malcolm, *Gaelic-English, English-Gaelic Dictionary* (Edinburgh: Mercat Press, 1979).

4. MacKillop, James, *Dictionary of Celtic Mythology (Oxford:* Oxford University Press, 1979).

5. Carmichael, Alexander (ed.) Matheson, Angus, *Carmina Gadelica: Hymns and Incantations*, Vol. 5, 1903.

6. Dames, Michael, *Mythic Ireland* (London: Thames and Hudson, 1992) pp. 247-257.

7. Ibid., p. 250.

8. Abercrombie J., The Wandering Psyche, *Folklore*, Vol. 6 (1895) pp. 165-7.

9. Dames, Michael, *The Avebury Cycle* (London: Thames and Hudson, 1977) pp. 85-6.

10. Mythic Ireland, op. cit., pp. 252-3.

11. Carmichael, A., *Carmina Gadelica*, Vol. 1, p. 143.

12. Mythic Ireland, op. cit., pp. 248-9.

13. Carmichael, A., T*he Sun Dances: Prayers and Blessings from the Gaelic* (Edinburgh: Christian Community Press, 1954) pp. 49-51.

14. Allen, J. Romilly, The Archaeology of Lighting Appliances, *Proceedings of the Society of Scottish Antiquaries* (1888) p. 82, figs.1-3.

15. *Carmina Gadelica*, Vol. 1, op. cit., p. 143.

16. Kennedy-Fraser, Marjory, quoted in Atheling, William, *The New Age*, Vol.12 (April 1919) p. 16.

17. Gordon Childe, Vere, et. al., *Skara Brae: A Pictish Village in Orkney* (Orkney: Kegan Paul, Trench, Trubner & Company, 1931).

18. Ibid.

19. The Orkney Venus, *British Archaeology*, Vol. 113 (2010) p. 8.

20. *Mythic Ireland*, op. cit., p. 246.

21. Sands of Time: Domestic Rituals at the Links of Noltland, *Current Archaeology*, Number 272 (October 2012).

22. Calder, C.S.T., *A Neolithic Temple at Stanydale, Shetland*, 1950, in Daniel, G., *The Megalith Builders of Western Europe* (London: Hutchinson,1958). Calder, C.S.T., 'Report on the Excavation of a Neolithic Temple at Stanydale in the Parish of Sandsting, Shetland', *Proceedings of the Society of Antiquaries of Scotland*, Volume 84 (1949-50) pp. 185–205.

23. Fiona Hyslop, Scottish Cabinet Secretary for Cultural and External Affairs, in 2012.

24. Eliade, Mircea, quoted in Stoller, P., Multiple Identity Enactments, *Anthropology and Humanism Quarterly*, 1986.

25. Ritchie, Anna, Excavation of Pictish and Viking-age Farmsteads at Buckquoy, Orkney, *Proceedings of the Society of Scottish Antiquaries*, Vol. 108 (1976-7) pp. 174-8.

26. Jackson, Anthony, quoted in the above report, p. 183.

27. Thomas, Charles, The Animal Art of the Scottish Iron Age and its Origins, *The Archaeological Journal*, Vol. 118 (1961) p. 14-63.

28. Kennedy-Fraser, op. cit.

29. Carmina Gadelica, Vol. 1, op. cit., p. 259.

30. Ibid., p. 247, and p. 281.

31. Jackson, Laura (Riding), *The Word 'Woman' and other Related Writings* (Manchester: Carcanet, 1994).

• **The Great British Coastal Walk**
The places and items mentioned are referenced from contemporary editions of the Ordnance Survey Landranger and Explorer Maps of Great Britain, named in a sequence heading east around the island, starting from the Scilly Isles and Land's End in Cornwall. (Sheets 101-451; plus OL series, sheets 15, and 20, 23, 27, 29 and 35-6.)

O.S. Gazetteer of British Place Names (Cardiff: Association of British Counties, 1975).

Smith, A.H. (ed.) *English Place-Name Elements,* 2 Parts (Cambridge: Cambridge University Press, 1956-8).

Macdonald, James, *Place names of West Aberdeenshire* (Aberdeen: Aberdeen University, 1900).

Johnston, James B., *Place-Names of Scotland* (London: John Murray, 1934).

Watson, William J., *The History of the Celtic Place-names of Scotland* (London William Blackwood & Sons, 1926).

Dwelly, Edward, The Illustrated Gaelic-English Dictionary (10th ed.) (Edinburgh: Birlinn Limited, 2001).

Robertson, Boyd, and Macdonald, Ian, *Teach Yourself: Essential Gaelic Dictionary* (London: Hodder Education, Teach Yourself, 2010).

Collins Spurrell, Pocket *Welsh Dictionary*, 4th Revised Edition (2009).

Ekwall, Eilert, *English River Names* (London: OUP, 1928).

• **Silbury's Missing Myths**
1. Ricoeur, P., in *The Encyclopaedia of Religion*, Vol. 10 (New York: Macmillan Publishing Company, 1987) pp. 261-68.

2. Dames, Michael, *The Silbury Treasure: The Great Goddess Rediscovered* (London: Thames and Hudson, 1976) pp. 104-128.

3. Dames, Michael, *The Avebury Cycle* (London: Thames and Hudson 1977).

4. Ibid., quoted p. 11.

5. Leary, Jim, in *British Archaeology*, Issue 70 (May 2003), p. 17.

6. MacNeill, Máire, *The Festival of Lughnasa* (London: OUP, 1962) pp. 312-38.

7. Gimbutas, Marija, *The Goddesses and Gods of Old Europe* (Oakland, California: University of California Press, 1974). And Preston, J.J., Goddess Worship, an Overview, in *Encyclopaedia of Religion*, Vol. 6 (New York: Macmillan Publishing Company, 1987) pp. 35-44.

8. MacKillop, James, *A Dictionary of Celtic Mythology* (Oxford: OUP, 1998).

9. Quarrie, Deanne (Createspace, *Dancing with Goddess*, 2010).

10. MacNeill, op. cit., p. 10.

11. Mac Cana, Proinsias, *The Cult of the Sacred Centre* (Dublin: 2011) pp. 258-61.

12. Dames, Michael, *The Silbury Treasure*, pp. 166-174.

13. Stukeley, William, *Abury, A Temple of the British Druids* (London: 1743), p. 43. and Long, W., in *Wiltshire Archaeology and Natural History Society*, Vol. 4 (1874).

14. MacNeill, op. cit., pp. 38-9.

15. Buckley, J., in *Wiltshire Archaeology and Natural History Society*, Vol. 11 (1857) pp. 149-162.

16. *Wiltshire Archaeology and Natural History Society*, Vol. 56 (1957) p. 124.

17. Grigson, G., *The Englishman's Flora* (London, J.M. Dent, 1987).

18. Cuan O Lothcain, In Praise of the Tailtiu Oenach, 1007 A.D.

19. Smyth, Daragh, *A Guide to Irish Mythology* (Dublin: Irish Academic Press, 1988) pp. 49-52.

20. Wilde, Sir W., *Beauties of the Boyne and the Blackwater*, 2nd edition (1850).

21. Hull E., *Folklore of the British Isles* (London, 1928) pp. 23-4.

22. Reports by Murphy, M.J., July 17th-31st, (1948), quoted in MacNeill, op. cit., p.160-61.

23. Pepys, S., *Diary*, Vol. 3 (London: 1854) p. 466.

24. Mac Cana, op. cit., pp. 256-61.

25. Palm (Fig) Sunday harvest celebrations reported in *Wiltshire Archaeology and Natural History Society*, Volume 37 (1911-12) p. 483. See also the Bath Journal of August 24, 1746; and WANHS Vol 49 (1940-42) p. 126, for a reference to children gathering on Silbury, as a relict of the Palm/Fig Sunday ritual, quoting Fred S. Thacker's *The Strippling Thames* (Oxford: Blackwell, 1932).

26. Ibid.

27. *John Barleycorn*, version by Burns, R., from Ballantyne MS, of 1538.

28. Long, R., *A Silbury Hill Walk*, 1970, displayed in the Hayward Gallery, London, 1991.

29. Whitehead, Paul, and Edmunds Mike, *Silbury Hill, Wiltshire: Palaeohydrology of the Kennet, Swallowhead Springs and the Siting of Silbury Hill*, Environmental Studies Report, No.12-2012 (Portsmouth: English Heritage, 2012).

30. Nietzsche, Friedrich, *The Gay Science*, para. 341, 1882.

31. Mac Cana, op. cit., p. 261.

32. Dames Michael, *The Avebury Cycle* (London: Thames and Hudson, 1977).

33. Pass A.C. Recent Excavations at Silbury Hill, in *The Clifton Antiquarian Club*, Vol. 1 (1887) pp. 130-135; and Pass, A.C., WANHS, Vol. 23 (1887) p. 246. Pass sank ten vertical shafts into the Silbury ditch in 1886, and found that beneath a foot of modern soil, lay 16 feet of white alluvial clay, that had accumulated over more than 2,000 years. The lowest layers contained worked Neolithic flints. The chalk bedrock lay at between 16 and 21 feet below the modern surface. No such alluvial silt was found,

between the two natural chalk causeways on the Hill's south side, into which the Goddess's August 1st 'corn baby' was delivered, indicating that this section of the ditch had always been dry.

34. Valentin, J., *Silbury Moat, Roman Coins,* in WAHNS, Vol. 94, 2001, p. 162.

35. Field, D. in *British Archaeology,* Vol. 70 (May 2003) p. 17.

36. Leary J., and Field, D., Great Monuments, Great Rivers, in *British Archaeology,* Issue 61, Sept. 2001 pp. 27-31.

37. Marshall, D., Silbury Springs, in *British Archaeology,* Issue 131 (July 2003) pp. 24-7.

38. Dames, Michael, *The Silbury Treasure: The Great Goddess Rediscovered* (London, Thames and Hudson, 1976) pp. 41-66.

39. *Swallowhead Springs: An Ancient Sacred Site Reborn* (London: Pixyled publications 2014).

40. Whitehead, Paul, and Edmunds, Mike, *Silbury Hill, Wiltshire: Palaeohydrology of the Kennet, Swallowhead Springs and the Siting of Silbury Hill,* Environmental Studies Report, No.12-2012 (Portsmouth: English Heritage, 2012).

41. Field, David and Leary, Jim, *The Story of Silbury Hill* (London: English Heritage, 2010).

42. Atkinson, R.J.C., in *The Times Educational Supplement* (October, 1976).

43. Plato's dualism, in Speake, Jennifer (ed.) *A Dictionary of Philosophy* (London: Pan Reference, 1979) pp. 249-252.

VIII - THE CHRISTIAN OVERLAY

• Devil Talk

1. Dames, Michael, *The Avebury Cycle* (London: Thames & Hudson, 1977 and 2nd edition 1996).

2. Milton, J. *Paradise Lost* Book 1, lines 283-313.

3. *Shorter Oxford English Dictionary,* Sixth edition (Oxford University Press, 2007).

4. Wiltshire, Kathleen and Carrott, Patricia M.C. (eds) *Wiltshire Folklore* (Salisbury: Compton Press, 1975).

5. Field, David, *Silbury Hill Survey Report* (English Heritage, 2002) p. 60.

6. *The Ordnance Survey Gazetteer of Great Britain,* 3rd Edition (London: Palgrave Macmillan, 1990).

7. Robertson, Dawn and Koronka, Peter, *Secrets and Legends of Old Westmorland* (Kirkby Stephen, Cumbria: Pagan Press, 1992).

8. Clwyd Paranormal Database (www.paranormaldatabase.com/wales/clwyd). At Pen y Cefn mount.

9. Ibid., The Devil's Tree at Llanrhos.

10. Horovitz, David, *The Place-Names of Staffordshire* (Nottingham: Brewood, Stafford, 2009).

11. Anon, 1813. Taken from verses intended to be cast onto the bells of Dudley's new church.

• Sheela-na-gig

1. Andersen, Jorgen, *The Witch on the Wall: Medieval Erotic Sculpture in the British Isles* (Denmark: Rosenkilde and Bagger, 1977).

2. Weir, Anthony and Jerman, James, *Images of Lust: Sexual Carvings on Medieval Churches* (London: Routledge, 2013).

3. Roberts, Jack, *Sheela-na-Gigs, White Dragon* (August 1995) p. 4.

4. McMahon, Joanne and Roberts, Jack, *The Sheela-na-Gigs of Britain and Ireland: the Divine Hags of the Christian Celts: an Illustrated Guide. (Cork: Mercier Press,* 2001).

5. Weir and Jerman, op. cit. p. 23.

6. *The Irish Times,* 23rd September, 1997.

7. *Murray, M.A., Female Fertility Figures,* The Journal of the Royal Anthropological Institute of Great Britain and Ireland, Vol. 64 (1934) pp. 93-100.

8. Kelly, Eamonn P., *Sheela-na-gigs: Origins and Functions (Irish Treasure Series)* (Dublin: Town House, 1996).

9. Andersen, op. cit.

10. The title Sheela-na-Gig is regarded as a spurious invention by Barbara Freitag, see *her Sheela-Na-Gigs: Unravelling an Enigma* (London: Routledge, 2004) and by several other writers, including Lawlor, H.C., in *Man, Royal Anthropological Institute of Great Britain and Ireland,* Vol. 31 (January, 1931).

11. Dames, Michael, *The Silbury Treasure: the Great Goddess Rediscovered* (London: Thames and Hudson, 1976).

12. Freitag, op. cit.

13. Squatting, on her haunches, near to the ground.

14. Eliade, Mircea, 'Towers' in *Encyclopedia of Religion* Vol. 14 (1993) p. 584.

15. Dames Michael, *The Avebury Cycle* (London: Thames and Hudson, 1977), p. 103 (see incisions on the body of the goddess Nerthus).

16. *Shorter Oxford English Dictionary,* 6th edition (Oxford: OUP, 1997).

17. Worsley, Richard, *The History of the Isle of Wight* (A. Hamilton, 1781), quoted by Andersen, op. cit., p. 11.

• Mary Madonna

1. Daly, Mary, *Beyond God the Father: Towards a Philosophy of Women's Liberation* (Boston, Beacon Press, 1973).

2. Mauricius Flavius, Emperor of Constantinople, c. 1600.

3. Bishop of Livia on the Jordan, c. A.D. 600.

4. Council of Ephesus, A.D. 431.

5. Council of Ephesus doctrine, A.D. 431

6. Pope John Paul II, 1988.

7. Mulack, Christa, *Mary – the Secret Goddess in Christianity* (Stuttgart: 1985).

8. Telesco, Patricia, *365 Goddess: a Daily Guide to the Magic and Inspiration of the Goddess* (San Francisco: Harper San Francisco, 1999).

9. Servasanctus, *Exordium,* Ch. xix; *and 150 Praises to Mary,* c. 1300.

10. Marracci, Hippolytus, *Polyanthea Mariana,* c. 1612, Rome.

11. Hopkins, Gerard Manley, The Blessed Virgin Compared to the Air We Breathe, *Collected Poems,* (1883), ll. p.p.34-8, p.p.113-7, and p.p.124-6. See also Pick, John (ed) *A Hopkins Reader,* (Oxford: OUP, 1953).

• **Kathy Jones: Individual and Deity**
1. Eitel, E. and Michell, J., *Feng Shui: the Science of Sacred Landscape in Old China* (New Mexico, USA: Synergetic Press, 1993).
Baldrain, F., Taoism (transl.) le Blanc, C., in Eliade, Mircea, *Encyclopedia of Religion*, Vol. 14 (Detroit, USA: Macmillan, 1987) pp. 290-293.
Naofusi, H., *Shinto*, in *Encyclopedia of Religion*, Vol. 13, pp. 280-81.
2. Townsend, R.F., Sacred Geography, in *Encyclopedia of Religion*, Vol. 5, pp. 509-512. Eck, D.L., 'Mountains' in *Encyclopedia of Religion*, Vol. 12, pp. 425-7. Eck, D. L., 'Rivers', in *Encyclopedia of Religion*, Vol. 12, pp. 425-7.
3. Berndt, R.M., Australian Religions, an overview, in *Encyclopedia of Religion*, Vol. 1, pp. 530-547.

Index

Index

Statius vi
Stukeley, William 104
sublime, the 42, 130, 166
sùil 169, 186, 295
Sulcus Primigenius 43
supernatural 3, 5, 8-9, 16-17, 32-35, 41, 45, 56, 67, 69-71, 74, 81, 84, 94, 101, 107-108, 113, 120, 122, 128, 135-136, 140, 183, 186, 189, 191-192, 204-207, 220, 230, 233-234, 240, 250-251, 255, 263, 272-273, 275, 280, 283, 296, 309
superstition 104
Swallowhead springs 41, 45, 51, 72, 80-81, 101-106, 121, 154, 158-160, 222, 274, 277, 285-287
swine 9, 117

Taliesin 1, 3, 62, 119
Tellus vi, 87, 155, 157
temple vi, 17-18, 34, 44, 46-47, 84, 87-88, 90, 92, 131, 139, 153, 156, 161-171, 184, 186, 190-191, 206, 211, 246, 256-257, 306-307, 309
Thames (river) 51, 80, 102, 104, 110, 112, 165, 180, 267
The Gop (earthwork) 136
Thor's Well 105
threshold 88, 207, 236-237
Tolstoy, Leo 46
tomb 136, 151-152
tongue 6, 206, 278
topography v, 36-37, 40, 189, 208-209, 261
treasure 12, 22, 48
Triads 119, 245, 250
Triple Goddess 21, 84, 94, 103, 235
Triptolemus 69, 158
Turning Castle 119
Tyche 168, 209

Uffington 177-183
Uisneach 203-204, 208, 183-209, 219, 279
Ulster 6, 183, 199-200
underworld v, 3, 8, 35, 46-47, 50-52, 57, 62, 69, 72-74, 87, 95, 99, 101, 103-104, 110, 118, 124, 126, 132, 145, 158-159, 167-168, 178, 187, 203, 207-208, 219,

221, 229, 238, 243, 250, 278, 283-286, 291, 295, 301
Universal Mother 49
Uranus vi-vii, 87
Uriconium 123
Urien Rheged 120
Urisk (half-goat, half-human water spirit) 113

vagina 104, 254
Venus 7, 24, 166
Verbeia (goddess) 166
Verstehen 227
Virgil 157
Virgin Mary 8, 25, 125, 211, 252, 255, 296, 298
voices (anomalous) 120, 302

Waden Hill 286
Wales 1-3, 10, 14, 18, 22-24, 60, 62, 64, 69-70, 79, 83, 93, 95, 106-108, 112, 114, 118-121, 134, 137, 165, 173, 180, 182, 198-199, 220, 222, 230-232, 234, 236, 240, 242, 244-245, 250-251, 263, 265, 270, 272, 292
water 3, 12-13, 16, 21, 23, 27, 34, 38, 42, 45, 47-51, 55-56, 64, 72, 78, 80, 90, 99, 101-117, 124-125, 134, 137, 144, 153, 155-156, 159, 169, 171-173, 187, 199, 222, 237, 242-243, 247-248, 250, 262, 271, 276, 285-287, 298 (body) 291, 295 (table) 283 (sprites) 113-114, 270-271 (goddesses) 110, 155, 231, 286
Water Leapers (Welsh water spirit) 114
Water Wraiths 113
Wayland the Smith 182
wedding 18, 71, 77, 92, 116, 166, 292, 294 (chamber) 202
well-dressing 104-105
Welsh giants 93-95
West Kennet Long Barrow 44, 51, 52, 159, 285
wheat 2, 46, 62-64, 72, 79, 125, 155-157, 198, 240, 277, 284
wheel 12, 131-132, 173, 177-179, 186, 191, 198, 215, 264, 292
Whistler, the 57

349